WAR AND ANTI-WAR

By the same authors

Future Shock
The Third Wave
Powershift
The Adaptive Corporation
Previews and Premises
The Eco-Spasm Report
The Culture Consumers

WAR AND ANTI-WAR

SURVIVAL AT THE DAWN OF THE 21ST CENTURY

ALVIN AND HEIDI TOFFLER

LITTLE, BROWN AND COMPANY

BOSTON   NEW YORK   TORONTO   LONDON

First Edition

Library of Congress Cataloging-in-Publication Data
Toffler, Alvin.
    War and anti-war : survival at the dawn of the twenty-first
century / by Alvin and Heidi Toffler. — 1st ed.
        p.   cm.
    Includes bibliographical references and index.
    ISBN 0-316-85024-1
    1. Military art and science — Economic aspects. I. Toffler,
Heidi. II. Title.
U102.T64   1993
355.02 — dc20   93-20789
                                     CIP

10 9 8 7 6 5 4 3 2 1

RRD-VA

Published simultaneously in Canada
by Little, Brown & Company (Canada) Limited

Printed in the United States of America

For Betty and Karen

You may not be interested in war,
but war is interested in you.

—*Trotsky*

CONTENTS

WAR AND ANTI-WAR

INTRODUCTION

THIS BOOK is about wars and anti-wars to come. It is for the Bosnian child whose face has been half ripped away by explosives, and for his mother staring with glazed eyes at what is left. It is for all the innocents of tomorrow who will both kill and die for reasons they do not understand. It is a book about peace. Which means it is a book about war in the startling new conditions we are creating as we race together into an alien future.

A fresh century now stretches before us, one in which vast numbers of humans can be raised from the edge of hunger . . . in which the ravages of industrial-era pollution can be reversed and a cleaner technology created to serve humanity . . . in which a richer diversity of cultures and peoples can participate in shaping the future . . . in which the plague of war is stanched.

But we appear, instead, to be plunging into a new dark age of tribal hate, planetary desolation, and wars multiplied by wars. How we deal with this threat of explosive violence will, to a considerable extent, determine how our children live or, perhaps, for that matter, die.

Yet many of our intellectual weapons for peacemaking are hopelessly out of date — as are many armies. The difference is that armies all over the world are racing to meet the realities of the twenty-first century. Peacemaking, by contrast, plods along, trying to apply methods more appropriate to a distant past.

The thesis of this book is clear — but as yet little understood: the way we make war reflects the way we make wealth — and the way we make anti-war must reflect the way we make war.

No subject is as easily ignored by those of us lucky enough to be living in peace. After all, we each have our private wars for survival: making a living, caring for our family, battling an illness. Enough, it would seem, to worry about these immediate realities. Yet how we fight our personal, peacetime wars, how we live our daily lives, is deeply influenced by real, and even by imagined, wars of the present, past, or future.

Present-day wars raise or lower the price of gasoline at the pump, food in the supermarket, shares on the stock exchange. They ravage the ecology. They erupt into our living rooms via our video screens.

Past wars reach across time to affect our lives today. The torrents of blood spilled centuries ago over issues now forgotten, the bodies charred, impaled, broken, or blown into nothingness, the children reduced to swollen bellies and stick-limbs — all shaped the world we inhabit today. To cite a single, little-noticed example, wars fought a thousand years ago led to the invention of chain-of-command hierarchies — a form of authority familiar to millions of jobholders today. Even wars of the future — whether planned or merely imagined — can steal our tax dollars today.

Not surprisingly, imagined wars grip our minds. Knights, samurai warriors, janissaries, hussars, generals, and G.I. Joes parade relentlessly through the pages of history and the corridors of our mind. Literature, painting, sculpture, and movies picture the horrors, heroism, or moral dilemmas of war, real and unreal.

But while wars actual, potential, and vicarious shape our existence, there is a completely forgotten reverse reality. For every one of our lives has also been shaped by wars that were NOT fought, that were prevented because "anti-wars" were won.

War and anti-war, however, are not either/or opposites. Anti-wars are not just waged with speeches, prayers, demonstrations, marches, and picket lines calling for peace. Anti-wars, more importantly, include actions taken by politicians, and even by warriors themselves, to create conditions that deter or limit the extent of war. In a complex world, there are times when war itself becomes an instrument needed to prevent a bigger, more terrible war. War *as* anti-war.

At the highest level, anti-wars involve strategic applications of military, economic, and informational power to reduce the violence so often associated with change on the world stage.

Today, as the world hurtles out of the industrial age and into a new century, much of what we know about both war and anti-war is dangerously out of date. A revolutionary new economy is arising based on knowledge, rather than conventional raw materials and physical labor. This remarkable change in the world economy is bringing with it a parallel revolution in the nature of warfare.

Our purpose therefore is not to moralize about the hatefulness of war. Some readers may confuse the absence of moralizing for an absence of empathy with the victims of war. This is to assume that cries of pain and anger are enough to prevent violence. Surely there are enough cries of pain and enough anger in the world. If they were sufficient to produce peace, our problems would be over. What is missing is not more emotive expression but a fresh understanding of the relations between war and a fast-changing society.

This new insight, we believe, could provide a better base for action by the world community. Not crash-brigade, after-the-fact intervention, but future-conscious preventive action based on an understanding of the shape that wars of tomorrow may assume. We offer here no panacea. What we offer, instead, is a new way of thinking about war. And that, we believe, may be a modest contribution to peace, for a revolution in warfare requires a revolution in peacefare as well.

Anti-wars must match the wars they are intended to prevent.

PART ONE

CONFLICT

1

UNEXPECTED ENCOUNTER

THE TRAIL started with an unexpected phone call, a night-time meeting in a motel near Washington, and a U.S. Army general in civilian clothes. We had not met him before, and didn't know why he wanted to see us. We had no intention of writing these pages.

At 7:30 P.M. on April 12, 1982, a short, slight, black-browed man strode out of the elevator of the Quality Inn near the Pentagon and joined us. Don Morelli introduced himself. Born of an immigrant Italian family in Pennsylvania, he was a West Pointer who had led combat troops in the Mekong Delta in Vietnam. But, as we were soon to discover, the most important battle of his life was yet to come.

It is often charged that the military brass spend their time preparing to fight the last war over again. From Don Morelli that night we learned that the same charge can be hurled at the intellectuals, politicians, and protesters who claim to speak for peace. The fact is, much of what is now publicly said or written about both war and peace is obsolete. It was conceived in Cold War categories and, worse yet, frozen in the mind-set of the smokestack era.

Don Morelli began his conversation with the news that a group of American generals were busy reading our 1980 book, *The Third Wave*. That book argued that the agricultural revolution of 10,000 years ago launched the first wave of transformatory change in human history; that the industrial revolution of 300 years ago triggered a second wave of change; and that we, today, are feeling the impact of a third wave of change.

Each wave of change brought with it a new kind of civilization. Today, our book suggested, we are in the process of inventing a revolutionary Third Wave civilization with its own economy, its own family forms, media, and politics.

That work, however, said almost nothing about war. Why, then, we wanted to know, were our generals under instruction to study it?

BRUTE FORCE TO BRAIN FORCE

The reason, Morelli explained, was that the same forces transforming our economy and society were about to transform war as well. Almost unknown to the outside world, a group had been put in place to design the revolutionary military of the future.

He told us that this team, led by his boss, a Kansas-born general named Donn A. Starry, had set out to reconceptualize war in "Third Wave" terms, to train soldiers to use their minds and fight in a new way, and to define the weapons they would need. Morelli's job was "doctrine." His task was to formulate, in effect, military doctrine for a Third Wave world.

We spoke for hours. We spoke about everything from video games to corporate decentralization, from the frontiers of technology to the philosophy of time. All these and more, he said, were involved in the reconceptualization of war.

After dinner Morelli took us upstairs to his room, where he had two slide projectors set up. It was the same briefing he had previously given to George Bush, then vice president of the United States. The hours sped by as we looked at slides and fired questions at him.

This was, it pays to recall, almost ten years before the term "smart bomb" became part of the world's vocabulary. The U.S. military was still demoralized by its defeat in Vietnam. But Morelli's mind was on the future, not the past, and what we saw in that room was an amazing preview of what the entire world watched breathlessly on CNN a decade later during the Gulf War.

In fact, what we saw pointed in directions not understood even today by the world public, a transformation of military power that can only be understood as we uncover, in the chapters that lie ahead, the remarkable parallels between the emerging economy of the future and the fast-changing nature of war itself, each accelerating change in the other.

Put simply: as we transition from brute-force to brain-force economies, we also necessarily invent what can only be called "brain-force war."

Don Morelli showered us with striking ideas. The American military's biggest problem? It let technology drive strategy, rather than letting strategy determine technology. The most important change in war since Vietnam? Precision-guided weapons. The biggest problem for democracies in relation to the military? Democratic armies cannot win wars without popular support, a consensus behind them. But crises could now arise faster than consensus could form. Can nuclear war be avoided? Yes. But not in an orthodox way. Why was he interested in the passages we had written about the philosophy of time? Because the military had to shift from an orientation toward space to an orientation toward time. Morelli now wound up his sparkling intellectual performance.

Psychiatrists call the last few words spoken by a patient after a therapy session "leakage." And, they say, the leakage often is more important than all the rest of the hour. As we stood in the doorway trying to make sense of what we had heard, Morelli dropped his own personal bombshell.

"I'm forty-nine years old," Morelli now confided, "and I'm dying of cancer." He paused.

Then with a finality that expressed long and careful self-examination, Morelli declared, "I will consider it the fulfillment of my life's mission if the new doctrine I've outlined for you tonight is actually implemented by the United States and our allies."

For good or ill — or both — Morelli's life mission has been more than fulfilled.

BEYOND THE COMIC STRIP

That first meeting led to others in Washington and at Fort Monroe, in Virginia. The Don Morelli we came to know did not fit anyone's stereotype of the soldier. Intellectuals, in particular, have tended to caricature military men as brutish or just plain stupid. Think of political cartoons picturing pigeon-breasted generals dripping with medals and sashes, their faces devoid of intelligence. Think of Gilbert and Sullivan's satirical song "I Am the Very Model of a Modern Major-General," or the First Lord of the Admiralty in *H.M.S. Pinafore*, who claimed, "I thought so little, they rewarded me / By making me the Ruler of the Queen's Navee!"

Whatever basis in reality such comic-strip images may once have had, and may still have in some other countries, they did not apply to Don Morelli or the officers to whom he subsequently introduced us. Morelli was, in fact, an intellectual who wore a uniform (sometimes). An "up" personality, he was in love with ideas. He also radiated warmth, seeming to search not for the weakness in others, but for gentleness. He had a ready sense of humor, and never ran out of Italian jokes. He studied oil painting under another officer to whom he taught chess in return. He loved both classical music and Stan Getz. He was an execrable singer. And he read everything from science fiction to history and biography. Another American general whom we later met called him "our Renaissance Italian."

Don Morelli was a serious man in the most serious business of all, and he knew it. But he was fun to be with. He was a dying man, but he was alive.

The last time we saw him was poignant. He had invited us to Fort Monroe to meet his replacement. The reason was all too clear. That day in February 1984, after a lunch fixed by Patti, his wife, and shared with several officers in battle fatigues, Morelli accompanied us to a waiting car. We were alone for a moment.

"The doctors give me only two to six more months to live, and the army is getting ready to retire me. I treasure our acquaintanceship," he said, "and regret it won't have a chance to develop further." We told him that we, too, valued our times with him. At that, he opened the door of the motor-pool car and waved a final farewell as a sergeant drove us away.

Those encounters, first with Don Morelli, and later with Donn Starry and others, ultimately led us to a fresh understanding of the role played in human affairs by that most dramatic, tragic, and consequential of social processes: war.

If war was ever too important to be left to the generals, it is now too important to be left to the ignorant — whether they wear uniforms or not. The same applies, even more strongly, to anti-war.

2

THE END OF ECSTASY

INFORMED ADULTS, if asked what wars have taken place in the years since the end of World War II, would have little trouble ticking off the Korean War (1950–53), the Vietnam War (1957–75), the Arab-Israeli wars (1967, 1973, 1982), the Persian Gulf War (1990–91), and perhaps several others.

Few, however, would know that, depending on how we count, between 150 and 160 wars and civil conflicts have raged around the world since "peace" broke out in 1945. Or that an estimated 7,200,000 soldiers were slaughtered in the process. That is the figure for deaths alone — not for the wounded, tortured, or mutilated. Neither does it include the far larger number of civilians sacrificed. Or those who perished in the aftermath of combat.

Ironically, in all of World War I, the number of soldiers killed was only moderately larger: approximately 8,400,000. This means, amazingly enough, that in terms of combat deaths, even allowing for a wide margin for error, the world has fought almost the equivalent of World War I all over again since 1945.

When civilian deaths are added, the total reaches an astronomical 33 to 40 million — again, not counting the wounded, raped, dislocated, diseased, or impoverished.

People have shot, stabbed, bombed, gassed, and otherwise murdered one another in Burundi and Bolivia, Cyprus and Sri Lanka, Madagascar and Morocco. There are today nearly 200 members of the United Nations. War has been waged in well over sixty of the member countries. SIPRI, the Stockholm International Peace Research Institute, counted thirty-one armed conflicts in progress in 1990 alone.

In fact, in the 2,340 weeks that passed between 1945 and 1990, the earth enjoyed a grand total of only three that were truly war-free. To call the years from 1945 to the present the "postwar" era, therefore, is to compound tragedy with irony.

If we look back at all this horrendous brutality, we discover a distinct pattern.

A TRILLION-DOLLAR PREMIUM

It is now clear that the U.S.-Soviet nuclear stalemate of the past few decades actually served to stabilize the world after the 1950s. With countries divided into two sharply defined camps, each knew more or less where it fit in the global system. From the sixties on, the consequence of direct war between the nuclear superpowers was "mutually assured destruction." The result was that while hot wars might rage in Vietnam, Iran/Iraq, Cambodia, Angola, Ethiopia, or in even more remote Third World locations, they were not fought on the territory of the main powers and they were never central to the economic existence of those powers.

In recent years nearly a trillion dollars has been spent annually for military purposes, mainly by the superpowers and their allies. These vast sums can be thought of as "insurance premiums" paid by the major powers to keep hot wars from raging within their own borders.

The two superpowers, the United States and the former Soviet Union, clearly fueled certain wars by their clients, proxies, satellites, or allies, feeding them arms, assistance, and ideological ammunition. But, perhaps more often than not, they also served as stabilizing super-cops — suppressing conflicts among their dependencies, mediating or moderating local disputes, and generally keeping their camp followers in line because of the dangers of limitless nuclear escalation.

In 1983, in a book called *Previews and Premises*, we pointed out that someday our children would "look back on the great world struggle between capitalism and socialism with an amused, patronizing air — the way we now look back at the battle between the Guelphs and the Ghibellines" in the thirteenth and fourteenth centuries. Today the term "Cold War" already has a quaintly archaic ring. Since 1991 the Soviet Union has been a shattered memory and the two-sided military structure imposed on the world by the two nuclear superpowers has crumbled with it. What followed was extraordinary.

SLAVERY AND DUELING

The first response to this breakup of the Cold War framework was a bad case of collective ecstasy.

For almost half a century the doomsday clock ticked and the world held its breath. It is, therefore, easy to understand the mindless joy that greeted the end of the Cold War, as symbolized by the crash of the Berlin Wall. Normally sober politicians sang odes to the new era of peace supposedly upon us. Pundits wrote about "peace breaking out." A huge "peace dividend" awaited. Democracies, in particular, would never fight one another. Some thinkers even ventured the notion that war might soon join slavery and dueling in the museum of discarded irrationalities.

This was not the first such outbreak of runaway optimism. "Nothing," wrote H. G. Wells in 1914, "could have been more obvious to the people of the early twentieth century than the rapidity with which war was becoming impossible." Alas, it was not so obvious to the millions who shortly perished in the trenches of the First World War — the "war to end all wars."

Once that war was over, Pollyannaish prognostications once again filled the diplomatic air, and in 1922 the then-great powers solemnly agreed to sink many of their warships to slow down an arms race.

In 1928 Henry Ford announced that "people are becoming too intelligent ever to have another big war." In 1932 an enthusiasm for disarmament led the American president, Herbert Hoover, to speak of the need to reduce "the overwhelming burden of armament which now lies upon the toilers of the world." His objective, he said, was that "all tanks, chemical warfare and all large mobile guns . . . all bombing planes should be abolished."

Seven years later World War II, the most destructive war in history, erupted. When that war ended in 1945 with the atomic horrors of Hiroshima and Nagasaki, the United Nations was formed and once again the world basked briefly in the illusion that lasting peace was at hand — until the Cold War and the nuclear standoff began.

COMPETITION PULLS THE TRIGGER

In the wake of the Soviet implosion, predictions of lasting peace once again rang out and a new theory (actually an old one in new wrapping) suddenly became fashionable. A growing chorus of Western and especially American intellectuals began to argue that the shape of tomorrow would essentially be determined by economic, not military, warfare.

As early as 1986, in *The Rise of the Trading State*, Richard Rosecrance of the Center for International Relations at the University of California, Los Angeles, contended that nations were becoming so economically interdependent as to lessen their tendency to fight one another. Trade, not military might, was now the path to world power. In 1987 Paul Kennedy similarly counterposed economic and military strength in *The Rise and Fall of the Great Powers*. Kennedy stressed the dangers of "military overstretch."

Now strategist Edward Luttwak began arguing that military might would decrease in significance in a new era of "geo-economics." C. Fred Bergsten, director of the Washington-based Institute for International Economics, echoed the same theme, asserting the "primacy" of economic over security issues in the new global system. Economist Lester Thurow added his voice to the choir: "Replacing a military confrontation with an economic contest is a step forward," he writes. From now on, the real competition among countries will revolve around which one can make the best products, raise standards of living, and develop the "best-educated and best-skilled workforce."

Upbeat geo-economic theory was used as ammunition to help elect President Clinton to the U.S. presidency. If the theory was right, its advocates argued, the military budget could be slashed and overdue social programs financed without increasing the American government's huge deficit. Better yet, a Clinton administration could refocus America on domestic problems (his predecessor, Clinton charged, had devoted too much attention to foreign affairs). Moreover, if the real battlefield of tomorrow was the global economy, the United States needed an "Economic Security Council" to wage economic war.

In the face of today's blood-tinged headlines, the lemming chorus has quieted. Geo-economics began to look less and less persuasive as violence flared all around us. National political leaders, it turns out, are not bookkeepers. As in the past, the war-makers of the world do not merely calculate economic pluses and minuses before plunging into war. They calculate, instead, their chances of seizing, expanding, or retaining political power.

Even when careful economic calculation enters the picture, it is, as often as not, erroneous, misleading, and mixed with other factors. Wars have resulted from irrationality, miscalculation, xenophobia, fanaticism, religious extremism, and just plain bad luck when every "rational" economic indicator suggested that peace would have been a preferable policy for all.

Worse yet, geo-economic war is not a substitute for military conflict. It is, all too often, merely a prelude, if anything a provocation, to actual war, as it was in U.S.-Japanese economic rivalry leading up to the Japanese attack on Pearl Harbor in 1941. At least in that case, competition pulled the trigger.

Heartening though it may be, geo-economic reasoning is inadequate for two even more fundamental reasons. It is too simple and it is obsolete. Simple because it tries to explain world power in terms of only two factors — economic and military. Obsolete because it overlooks the growing role of knowledge — including science, technology, culture, religion, and values — which is now the core resource of all advanced economies and of military effectiveness as well. Thus the theory ignores what may be the most crucial factor of all in twenty-first-century world power. We are entering not the geo-economic era but the geo-information era.

For all these reasons it is no surprise that we now hear less and less about this bullet-riddled theory of geo-economics.

After the latest wave of collective ecstasy, the morning-after letdown came. The world looked as though it were about to break out in a rash of "local wars." But even now a dangerous misperception persists: the widely held notion that wars of the future, like those of the previous half century, will continue to be confined to small countries in more or less remote regions.

A typical statement came from no less a personage than a U.S. undersecretary of defense: "We have achieved in North America, Western Europe, and Japan a 'zone of peace' within which it is fair to say war is truly unthinkable." History, however, is studded with "unthinkable wars." Just ask the citizens of Sarajevo.

Perhaps because it is too horrible to contemplate, the public is still encouraged to discount the possibility of major wars inside the territory of the great powers themselves, or of local conflicts that drag in the major powers in spite of themselves. Yet the terrifying truth is that the era of marginalized murder, when all wars were fought by small states in faraway places, may be screeching to an end. If so, our most basic strategic assumptions will need revision.

3

A CLASH OF
CIVILIZATIONS

IT HAS BELATEDLY BEGUN to dawn on people that industrial civilization is coming to an end. Its unraveling — already evident when we wrote about the "general crisis of industrialism" in *Future Shock* (1970) — brings with it the threat of more, not fewer, wars — wars of a new type.

Today many use the term "postmodern" to describe whatever it is that comes after modernity. But when we spoke about this with Don Morelli and Donn Starry in the early 1980s, we referred instead to the differences between First Wave, or agrarian; Second Wave, or industrial; and now Third Wave armies.

Because massive changes in society cannot occur without conflict, we believe the metaphor of history as "waves" of change is more dynamic and revealing than talk about a transition to "postmodernism." Waves are dynamic. When waves crash in on one another, powerful crosscurrents are unleashed. When waves of history collide, whole civilizations clash. And that sheds light on much that otherwise seems senseless or random in today's world.

In fact, once we grasp the wave theory of conflict, it becomes apparent that the biggest shift of power now beginning on the planet is not between East and West or North and South, nor is it between different religious or ethnic groups. The deepest economic and strategic change of all is the coming division of the world into three distinct, differing, and potentially clashing civilizations.

First Wave civilization, as we've seen, was inescapably attached to the land. Whatever local form it may have taken, whatever language its people spoke, whatever its religion or belief system, it was a product of the agricultural revolution. Even today, multitudes live and die in premodern, agrarian societies, scrabbling at the unyielding soil as their ancestors did centuries ago.

Second Wave civilization's origins are in dispute. Some historians trace its roots to the Renaissance, or even earlier. But life did not fundamentally change for large numbers of people until, roughly speaking, three hundred years ago. That was when Newtonian science first arose. It is when the steam engine was first put to economic use and the first factories began to proliferate in Britain, France, and Italy. Peasants began moving into the cities. Daring new ideas began to circulate — the idea of progress; the odd doctrine of individual rights; the Rousseauian notion of a social contract; secularism; the separation of church and state; and the novel idea that leaders should be chosen by popular will, not divine right.

Driving many of these changes was a new way of creating wealth — factory production. And before long many different elements came together to form a *system:* mass production, mass consumption, mass education, mass media all linked together and served by specialized institutions — schools, corporations, and political parties. Even family structure changed from the large, agrarian-style household in which several generations lived together to the small, stripped down nuclear family typical of industrial societies.

To the people actually experiencing these many changes, life must have seemed chaotic. Yet the changes were, in fact, all closely interrelated. They were merely steps toward the full development of what we came to call modernity — mass-industrial society, the civilization of the Second Wave.

This new civilization entered history with a roar in Western Europe, fiercely resisted at every step.

THE MASTER CONFLICT

In every industrializing country bitter, often bloody battles broke out between Second Wave industrial and commercial groups and First Wave landowners in alliance, very often, with the church (itself a great landowner). Masses of peasants were forced off the land to provide workers for the new "Satanic mills" and factories that multiplied over the landscape.

Strikes and rebellions, civil insurrections, border disputes, nationalist uprisings erupted as the war between First and Second Wave interests became the master conflict — the central tension from which other conflicts derived. This pattern was repeated in almost every industrializing country. In the United States it required a terrible Civil War for the industrial-commercial interests of the North to vanquish the agrarian elites of the South. Only a few years later, the Meiji Revolution broke out in Japan, and once more Second Wave modernizers triumphed over First Wave traditionalists.

The spread of Second Wave civilization, with its strange new way of making wealth, destabilized relationships between countries as well, creating power vacuums and power shifts. Industrialization led to the expansion of national markets and the accompanying ideology of nationalism. Wars of national unification swept Germany, Italy, and other countries. Uneven rates of development, competition for markets, the application of industrial techniques to arms production, all disturbed prior power balances and contributed to the wars that tore Europe and its neighbors apart in the mid- and late nineteenth century.

In fact, the center of gravity of the world power system began to migrate toward industrializing Europe and away from the Ottoman Empire and the Czar's feudal Russia. Modern civilization, the product of the great Second Wave of change, took root most rapidly on the northern shores of the great Atlantic Basin.

As the Atlantic powers industrialized, they needed markets and cheap raw materials from distant regions. The advanced Second Wave powers thus waged wars of colonial conquest and came to dominate the remaining First Wave states and tribal units all over Asia and Africa.

Thus, just as industrializing elites ultimately won the struggle for power inside their own countries, they also won the larger struggle for world power.

A BISECTED WORLD

It was the same master conflict again — Second Wave industrial powers versus First Wave agrarian powers — but this time on a global rather than domestic scale, and it was this struggle that basically determined the shape of the world until recent times. It set the frame within which most wars took place.

Tribal and territorial wars between different primitive and agricultural groups continued, as they had throughout previous millennia. But these were of limited importance, and often merely weakened both sides, making them easy prey for the colonizing forces of industrial civilization. This happened, for example, in southern Africa as Cecil Rhodes and his armed agents seized vast territories from tribal and agrarian groups busy fighting one another with primitive weapons. Elsewhere, too, many seemingly unconnected wars around the world were, in fact, expressions of the main global conflict not between competing states but competing civilizations.

Yet the very biggest and most murderous wars during the industrial age were intra-industrial — wars that pitted Second Wave nations like Germany and Britain against one another, as each one struggled for global dominance while keeping the world's First Wave populations in their subordinate place.

The ultimate result was a clear division. The industrial era bisected the world into a dominant and dominating Second Wave civilization and scores of sullen but subordinate First Wave colonies. Most of us grew up in this world, divided between First and Second Wave civilizations. And it was perfectly clear which one held power.

A TRISECTED WORLD

Today, the lineup of world civilizations is different. We are speeding toward a totally different structure of power that will create not a world cut in two but sharply divided into three contrasting and competing civilizations — the first still symbolized by the hoe; the second by the assembly line; and the third by the computer.

That term, "civilization," may sound pretentious, especially to American ears, but no other term is sufficiently all-embracing to include such varied matters as technology, family life, religion, culture, politics, business, hierarchy, leadership, values, sexual morality, and epistemology. Swift and radical changes are occurring in every one of these dimensions of society.

As a new civilization arrives, it touches the fundamental and the trivial alike. Thus today we see an enormous number of things that were inconceivable, unavailable, or socially disapproved of in the past — everything from heart transplants to Frisbees and yogurt franchises, from condos and consultants to contact lenses, from spacewalks to Game Boy cartridges, from Jews for Jesus to New Age worship, from laser surgeons to CNN, from ecological fundamentalism to chaos theory.

Change all these social, technological, and cultural elements at once and you create not just a transition but a transformation, not just society but the beginnings, at least, of a totally new civilization.

But to introduce a new civilization onto the planet and then expect peace and tranquility is the height of strategic naïveté. Each civilization has its own economic (and hence political and military) requirements.

In this trisected world the First Wave sector supplies agricultural and mineral resources, the Second Wave sector provides cheap labor and does the mass production, and a rapidly expanding Third Wave sector rises to dominance based on the new ways in which it creates and exploits knowledge.

Third Wave nations sell information and innovation, management, culture and pop culture, advanced technology, software, education, training, medical care, and financial and other services to the world. One of those services might well also turn out to be military protection based on its command of superior Third Wave forces. (That is, in effect, what the high-tech nations provided for Kuwait and Saudi Arabia in the Gulf War.)

DE-COUPLING THE POOR

In Third Wave, brain-based economies mass production (which could almost be considered the defining mark of industrial society) is already an outmoded form. De-massified production — short runs of highly customized products — is the cutting edge of manufacture. Services proliferate. Intangible assets like information become the key resource. Uneducated or unskilled workers are made jobless. Old industrial-style behemoths collapse of their own weight, the GMs and Bethlehem Steels that dominated the age of mass production face destruction. Labor unions in the mass-manufacturing sector shrink. The media are de-massified in parallel with production, and giant TV networks shrivel as new channels proliferate. The family system, too, becomes de-massified: the nuclear family, once the modern standard, becomes a minority form, while single-parent households, remarried couples, childless families, and live-alones proliferate.

Culture shifts from one in which standards are clearly defined and hierarchical to one in which ideas, images, symbols swirl in a maelstrom, and the individual plucks individual elements with which to form his or her own mosaic or collage. Existing values are challenged or ignored.

The entire structure of society, therefore, changes. The homogeneity of Second Wave society is replaced by the heterogeneity of Third Wave civilization.

In turn, the very complexity of the new system requires more and more information exchange among its units — companies, government agencies, hospitals, associations, other institutions, and individual people. This creates a ravenous need for computers, digital telecommunications, networks, and new media.

Simultaneously, the pace of technological change, transactions, and daily life speeds up. In fact, Third Wave economies operate at speeds so accelerated that their premodern suppliers can barely keep pace. Moreover, as information increasingly substitutes for bulk raw materials, labor, and other resources, Third Wave countries become less dependent on First Wave or Second Wave partners, except for markets. More and more they do business with each other. Eventually, their highly capitalized knowledge-based technology will take over many tasks now done by the cheap-labor countries and actually do them faster, better — and more cheaply.

Put differently, these changes threaten to slash many of the existing economic links between the rich economies and the poor.

Complete de-coupling is impossible, however, since it is not possible to stop pollution, disease, and immigration from penetrating the borders of the Third Wave countries. Nor can the rich nations survive if the poor wage ecological war on them by manipulating their environment in ways that damage everyone. For these reasons, tensions between the Third Wave civilization and the two older forms of civilization will continue to rise, and the new civilization will fight to establish global hegemony, just as Second Wave modernizers did with respect to the First Wave premodern societies in centuries past.

THE DUCK SOUP PHENOMENON

Once the concept of a clash of civilizations is grasped, it helps us make sense of many seemingly odd phenomena — today's flaring nationalisms, for example.

Nationalism is the ideology of the nation-state, which is a product of the industrial revolution. Thus, as First Wave, or agrarian, societies seek to start or complete their industrialization they demand the trappings of nationhood. Former Soviet republics like the Ukraine or Estonia or Georgia fiercely insist on self-determination, and demand yesterday's marks of modernity — the flags, armies, and currencies that defined the nation-state during the Second Wave, or industrial, era.

It is hard for many in the high-tech world to comprehend the motivations of ultra-nationalists. Their puffed-up patriotism strikes many as amusing. It calls to mind the land of Freedonia in the Marx Brothers' movie *Duck Soup*, which satirized the notion of national superiority as two fictional nations went to war against one another.

By contrast, it is incomprehensible to nationalists that some countries allow others to invade their supposedly sacred independence. Yet the "globalization" of business and finance required by the advancing Third Wave economies punctures the national "sovereignty" the new nationalists hold so dear.

POETS OF GLOBALISM

As economies are transformed by the Third Wave, they are compelled to surrender part of their sovereignty and to accept increasing economic and cultural intrusions from one another. The United States insists that Japan restructure its retail distribution system (thus threatening to wipe out an entire social class of small shopkeepers along with the culture and family structure they represent). In return, Japan insists that the United States put more money into savings, think long range, and restructure its education system. Such demands would have been deemed unacceptable invasions of sovereignty in the past.

Thus, while the poets and intellectuals of economically backward regions write national anthems, the poets and intellectuals of Third Wave states sing the virtues of a "borderless" world. The resulting collisions, reflecting the sharply differing needs of two radically different civilizations, could provoke some of the worst bloodshed in the years to come.

If today's redivision of the world from two into three parts seems less than obvious right now, it is simply because the transition from Second Wave brute-force economies to Third Wave brain-force economies is nowhere yet complete.

Even in the United States, Japan, and Europe, the domestic battle for control between Third and Second Wave elites is still not over. Important Second Wave institutions and sectors of production still remain, and Second Wave political lobbies still cling to power. A perfect measure of this was provided in the United States during the fading days of the Bush administration when the Congress passed an "infrastructure" bill providing $150 billion to refurbish the old Second Wave infrastructure of roads, highways, and bridges, but only $1 billion to help build an electronic supercomputer network for the country — part of the infrastructure of the Third Wave. Despite its support for the high-speed network, the Clinton administration changed that ratio hardly at all.

The "mix" of Second and Third Wave elements in each high-tech country gives each its own characteristic "formation." Nevertheless, the trajectories are clear. The global competitive race will be won by the countries that complete their Third Wave transformation with the least amount of domestic dislocation and unrest.

In the meantime, the historic change from a bisected to a trisected world could well trigger the deepest power struggles on the planet as each country tries to position itself in the emerging three-tiered power structure. Trisection sets the context in which most wars from now on will be fought. And those wars will be different from those most of us imagine.

PART TWO

TRAJECTORY

4

THE REVOLUTIONARY PREMISE

FOR ALL the conservativism of military institutions, there have always been innovators calling for revolutionary change. Don Morelli and the other officers charged with rethinking how an army must fight in tomorrow's world were part of a long military tradition. In fact, historians have filled the shelves of libraries with books about "revolutions in warfare."

All too often, however, the term has been applied too generously. For example, war is said to have been revolutionized when Alexander the Great defeated the Persians by combining "the infantry of the West with the cavalry of the East." Alternatively, the word "revolution" is often applied to technological changes — the introduction of gunpowder, for instance, or the airplane or the submarine.

Admittedly these produced profound changes in warfare. Surely they had enormous impact on subsequent history. Even so, they are what might be called sub-revolutions. They basically add new elements or create new combinations of old elements within an existing "game." A true revolution goes beyond that to change the game itself, including its rules, its equipment, the size and organization of the "teams," their training, doctrine, tactics, and just about everything else. It does this not in one "team" but in many simultaneously. Even more important, it changes the relationship of the game to society itself.

By this demanding measure, true military revolutions have occurred only twice before in history, and there are strong reasons to believe that the third revolution — the one now beginning — will be the deepest of all. For only within recent decades have some of the key parameters of warfare hit their final limits. These parameters are range, lethality, and speed.

Armies that could reach further, hit harder, and get there faster usually won, while the range-restricted, less well-armed, and slower armies lost. For this reason, a vast amount of human creative effort has been poured into extending the range, increasing the firepower, and accelerating the speed of weapons and of armies.

A DEADLY CONVERGENCE

Take range. Throughout history warmakers have tried to extend their reach. Writing about the war of the fourth century B.C., the historian Diodorus Siculus reported that the Greek general Iphicrates, fighting on behalf of the Persians against the Egyptians, "made his spears half as long again, and the length of swords almost doubled," thus extending the range of the weapons.

Ancient devices like catapults and ballistas could heave a ten-pound rock or ball a distance of 350 yards. The crossbow, used in China in 500 B.C. and common in Europe by 1100, gave a soldier a "standoff" weapon of seemingly enormous reach. (So horrible was this weapon that in 1139 Pope Innocent II tried to ban its use.) Arrows reached an extreme range of about 380 yards in the fourteenth and fifteenth centuries. Yet for all the experimentation with archery over the centuries, the furthest range of any arrow, as late as the nineteenth century, was 660 yards, achieved by the Turks. And in actual fighting, the maximum range of weapons was seldom attained.

By 1942, Alexander de Seversky in his visionary book *Victory Through Air Power* urged the United States to develop aircraft capable of flying 6,000 miles, then seemingly impossible. Today — even leaving aside the potentials for space-based weaponry — there is scarcely any point on the globe that cannot in theory be targeted by intercontinental ballistic missiles, aircraft carriers, submarines, re-fueled long-range bombers, or combinations of these and other weapons systems. For all practical purposes, the extension of range has reached its terrestrial limits.

As with range, so with speed. In June 1991 the U.S. Defense Department made public its Alpha chemical laser, capable of producing a million watts of power, as part of the development of an anti-missile system. The laser can, if targeted correctly, reach an enemy missile at the speed of light, presumed to be the fastest speed possible.

And, as to lethality — the sheer kill-capacity of conventional weapons has increased by five orders of magnitude from the beginning of the industrial revolution to today. This means that today's non-nuclear weaponry, on average, is 100,000 times more deadly than it was when steam engines and factories began to change our world. As to nukes, we need only contemplate the consequences of 100 or 1,000 Chernobyls to appreciate the awesome threat they pose. It is only within this last half century that planetary doomsday scenarios became a serious subject of discussion.

In short, three distinct lines of military development have converged explosively in our time. Range, speed, and lethality all reach their outer limits at about the same moment of history — the present half century. If nothing else, this fact alone would justify the term "revolution in warfare."

AFTER THE ENDGAME

But this fact is not all. For in 1957, a mere dozen years after the first nuclear weapon was completed, Sputnik, the world's first spacecraft, burst into the heavens, opening an entirely new region to military operations. Space has already transformed terrestrial military operations in terms of surveillance, communications, navigation, meteorology, and a hundred other things. No previous breakthrough, from the first use of the sea or the air as regimes for military action, can compare to the long-range implications of this event.

A few years later, in announcing the U.S. drive to place a man on the moon, President John F. Kennedy declared that while "no one can predict with certainty what the ultimate meaning will be of the mastery of space," it may well be that space will "hold the key to our future on earth."

These qualitative, indeed fantastic changes in the nature of war and the military all have come in a short thirty–forty year span, the very moment when the dominant civilization on earth — Second Wave, or industrial, society — began its terminal decay. They came during the endgame of the industrial era, and at approximately the time when a new type of economy and society began to take form. Even as some nations industrialize, a Third Wave or postindustrial civilization is springing up in the United States, Europe, and the Asia Pacific region.

And this helps explain why the military revolution that lies ahead will be far deeper than most commentators have so far imagined. A military revolution, in the fullest sense, occurs only when a new civilization arises to challenge the old, when an entire society transforms itself, forcing its armed services to change at every level simultaneously — from technology and culture to organization, strategy, tactics, training, doctrine, and logistics. When this happens, the relationship of the military to the economy and society is transformed, and the military balance of power on earth is shattered.

A revolution of this profundity has happened only rarely in history.

5

FIRST WAVE WAR

Throughout history, the way men and women make war has reflected the way they work.

Despite a romantic belief that life in the earliest tribal communities was harmonious and peaceful, violent battles certainly occurred among pre-agricultural, nomadic, and pastoral groups. In his book *The Evolution of War*, Maurice R. Davie wrote of the "incessant intergroup hostility in which so many primitive tribes" found themselves. These small groups fought to avenge killings, to abduct women, or for access to protein-rich game. But violence is not synonymous with war, and it was only later that conflict took on the true character of war as such — a bloody clash between organized states.

When the agricultural revolution launched the first great wave of change in human history, it led gradually to the formation of the earliest premodern societies. It gave rise to permanent settlements and many other social and political innovations. Among these, surely one of the most important was war itself.

Agriculture became the womb of war for two reasons. It enabled communities to produce and store an economic surplus worth fighting over. And it hastened the development of the state. Together these provided the preconditions for what we now call warfare.

Not all premodern wars, of course, had economic ends. The literature on the causes of war attributes it to everything from religious fanaticism to inborn aggressiveness in the species. Yet, in the words of the late Kenneth Boulding, a distinguished economist and peace activ-

ist, war is "quite distinct from mere banditry, raiding, and casual violence. . . . It requires . . . a surplus of food from agriculture collected
in one place and put at the disposal of the single authority."

RITES, MUSIC, AND FRIVOLITY

This link between war and the soil was perfectly clear to the strategists and warriors of the past. The great Lord Shang, writing in ancient China, prepared a manual for statesmen, much as Machiavelli
did 1,800 years later. In it, Shang declares, "The country depends on
agriculture and war for its peace."

Shang served the state of Ch'in from 359 to 338 B.C. Again and again
in his politico-military handbook he advises the ruler to keep the
people ignorant, to avoid rites, music, and any frivolity that might take
their minds off farming and warfare. "If he who administers a country is
able to develop the capacity of the soil to the full and to cause the people
to fight to the death, then fame and profit will jointly accrue."

When population is sparse, Shang urges the ruler to encourage the
in-migration of the soldiers of neighboring feudal lords. "Promise
them ten years free of military service and put them to work on the
land, thus freeing up the existing population to wage war."

Lord Shang's prescription for maintaining military discipline gives
the flavor of his thinking: "In battle five men are organized into a
squad; if one of them is killed, the other four are beheaded." On the
other hand, victorious officers are to be rewarded with grain, slaves, or
even "a tax-paying city of 300 families."

Lord Shang was roughly contemporary with Sun-tzu, whose *The
Art of War* became a military classic. In his introduction to a recent
edition of that work, Samuel B. Griffith writes, "During the Spring
and Autumn armies were small, inefficiently organized, usually in-
eptly led, poorly equipped, badly trained, and haphazardly supplied.
Many campaigns ended in disaster simply because the troops could
find nothing to eat. . . . Issues were ordinarily settled in a day. Of
course, cities were besieged and armies sometimes kept in the field for
protracted periods. But such operations were not normal."

A SEASONAL OCCUPATION

Centuries later across the world things were not terribly different in
ancient Greece, as far as food and agriculture were concerned. Output
in agrarian societies was so low and food surpluses still so small that

over 90 percent of all manpower was needed simply to work the land. The departure of a son for military service could mean an economic catastrophe for his family. Thus, according to historian Philip M. Taylor, when Greek fought Greek war was "a seasonal occupation, with the volunteer soldiers coming mainly from farms which needed no looking after during the winter months."

Getting back to the farm quickly was essential. "The harvest demands of the triad of Greek agriculture — the olive, the vine, and grain — left only a brief month or two in which these small farmers could find time to fight," writes the classical scholar Victor Hanson in *The Western Way of War.*

Greek soldiers were sometimes told to bring a three-day supply of food with them when they turned up for military duty. After that they were dependent on the countryside. According to historian John Keegan, in wars between city-states, "The worst damage one city could do to another, after the killing of its citizen-soldiers on the battlefield, was to devastate its agriculture." Centuries later, long after the ancient Greek city-states had been swallowed up by history, the story was still the same. Everywhere in First Wave societies, warfare was about agriculture.

As with any historical generalizations, there are notable exceptions to the idea that First Wave armies were poorly organized, equipped, and led. No one would regard the Roman legions in their heyday as an ad hoc, badly organized force. Yet Griffith's comment about the ragtag character of armies in Sun-tzu's era could apply equally well across much of human history and in other parts of the world as well.

This was especially true in decentralized agrarian societies where feudalism held sway. There the king typically had to rely on his nobles to supplement his troops for any important campaign. But his call on them was usually strictly limited. In his masterful study *Oriental Despotism*, historian Karl A. Wittfogel writes: "The sovereign of a feudal country did not possess a monopoly of military action. As a rule, he could mobilize his vassals for a limited period only, at first perhaps for three months and later for forty days, the holders of small fiefs often serving only for twenty or ten days, or even less."

What's more, the vassal usually did not deliver his full force to his sovereign, but called up only a fraction. Often even this fraction was under no obligation to continue fighting for the king if the war took them abroad. In short, the king had full control only over his own troops. The remainder of his forces was usually a patchwork of temporary units of dubious skill, equipment, and allegiance.

A European feudal lord who was attacked, writes Richard Shelly Hartigan in a history of the civilian in warfare, "could hold his vassals to their military obligations until the invader was repulsed; but a lord bent on offensive war could keep his men in the field for only forty days out of each year. . . ." Like the ancient Greeks and the Chinese, they were needed on the land.

AN ABSENCE OF PAYCHECKS

In most First Wave armies, moreover, soldier pay was irregular, usually in kind rather than money. (The money system was still rudimentary.) Not infrequently, as in ancient China, victorious generals were paid off with land, the central resource of the agrarian economy. Of course, officers did far better than ordinary soldiers. The historian Tacitus, describing the Roman army, quotes a soldier's complaint that after a lifetime of "blows, wounds, hard winters, plague-filled summers, horrible war, or miserable peace," a lowly legionnaire, on being mustered out, might be given little more than a parcel of swampy or mountainous land somewhere. In medieval Spain and as late as the early nineteenth century in South America, land was still being paid to warriors in lieu of money.

First Wave military units thus varied greatly in size, capability, morale, leadership quality, and training. Many were led by mercenary or even mutinous commanders. As was true in the economy, communications were primitive, and most orders were oral, rather than written. The army, like the economy itself, lived off the land.

Like tools for working the soil, weapons were unstandardized. Agrarian hand labor was mirrored in hand-to-hand combat. Despite limited use of standoff weapons such as slings, crossbows, catapults, and early artillery, for thousands of years the basic mode of warfare involved face-to-face killing, and soldiers were armed with weapons — pikes, swords, axes, lances, battering rams — dependent on human muscle power and designed for close combat.

In the famous Bayeux tapestry, William the Conqueror is shown wielding a club, and as late as 1650–1700, even senior military commanders were expected to participate in hand-to-hand killing. Historian Martin Van Creveld notes that Frederick the Great "was probably the first commander in chief regularly depicted as wearing a suit of linen rather than armor."

Economic and military conditions may have differed in what Wittfogel termed "hydraulic societies," where the need for huge irrigation projects led to the mass mobilization of labor, early bureaucratization, and more formalized and permanent military establishments. Even so, actual combat remained largely a personalized, face-to-face affair.

In brief, First Wave wars bore the unmistakable stamp of the First Wave agrarian economies that gave rise to them, not in technological terms alone but in organization, communication, logistics, administration, reward structures, leadership styles, and cultural assumptions.

Starting with the very invention of agriculture, every revolution in the system for creating wealth triggered a corresponding revolution in the system for making war.

6

SECOND WAVE WAR

THE INDUSTRIAL REVOLUTION launched the Second Wave of historical change. That "wave" transformed the way millions of people made a living. And war once more mirrored the changes in wealth creation and work.

Just as mass production was the core principle of industrial economies, mass destruction became the core principle of industrial-age warfare. It remains the hallmark of Second Wave war.

Starting with the late 1600s, when the steam engine was introduced to pump water out of British mines, when Newton transformed science, when Descartes rewrote philosophy, when factories began to dot the land, when industrial mass production began to replace peasant-based agriculture in the West, war, too, became progressively industrialized.

Mass production was paralleled by the *lévee en masse* — the conscription of mass armies paid by and loyal not to the local landowner, clan leader, or warlord, but to the modern nation-state. The draft was not new, but the idea of a whole nation in arms — *Aux armes citoyens!* — was a product of the French Revolution, which roughly marked the crisis of the old agrarian regime and the political rise of a modernizing bourgeoisie.

After 1792, writes Yale historian R. R. Palmer, a wave of innovation "revolutionized warfare, replacing the 'limited' war of the Old Regime with the 'unlimited' war of subsequent times. . . . War before the French Revolution was essentially a clash between rulers. Since that event it has become increasingly a clash between peoples." It increasingly became a clash of conscripted armies as well.

BAYONETS AND COTTON GINS

In the United States, it was not until 1862–63, during the Civil War — in which the industrializing North defeated the agrarian South — that the draft was imposed (by both sides). Similarly, in Japan, half a world away, the introduction of the draft came shortly after 1868, when the Meiji Revolution started that country on its path toward industrialization. There the feudal samurai warrior was replaced by the draftee soldier. After each war, as tensions eased and budgets were cut, armies might revert to volunteers once more, but in crisis mass conscription was common.

The most dramatic changes in war came from new standardized weaponry that was now produced by mass production methods. By 1798, in the new United States the inventor of the cotton gin, Eli Whitney, was asking for a government contract to "undertake to Manufacture ten or Fifteen Thousand Stand of Arms," each stand consisting of a musket, a bayonet, a ramrod, wiper, and screwdriver. Whitney offered as well to make cartridge boxes, pistols, and other items by using "machines for forging, rolling, floating, boreing, Grinding, Polishing etc."

This was an amazing proposition in its time. "Ten or fifteen thousand stand of arms!" write the historians Jeanette Mirsky and Allan Nevins, was "a notion as fantastic and improbable as aviation was before Kitty Hawk."

War accelerated the industrialization process itself by spreading, for example, the principle of interchangeable parts. This basic industrial innovation was quickly put to use turning out everything from handguns to the pulleys needed on sail-driven warships. In preindustrial Japan, too, some of the earliest, primitive mechanization was for the purpose of producing arms.

That other key industrial principle — standardization — was also soon applied not merely to the weapons themselves but to military training, organization, and doctrine as well.

The industrial transformation of war thus went far beyond technology. Temporary ragtag armies led by the nobility were replaced by standing armies led by professional officers trained in war academies. The French created the *état-major* system to give officers formal training for senior command. In 1875 Japan created its own military academy after studying the French. In 1881 the United States set up the School of Application of Infantry and Cavalry at Fort Leavenworth, Kansas.

A BARRAGE OF MEMOS

The division of labor in industry was reproduced in the rise of new specialized branches of the military. As in business, bureaucracy grew. Armies developed general staffs. Written orders replaced oral commands for many purposes. Memos proliferated in business and on the battlefield alike.

Everywhere industrial-style rationalization became the order of the day. Thus write Meirion and Susie Harries in *Soldiers of the Sun*, their impressive history of the Japanese imperial army, "The 1880s were the years when the army evolved and entrenched a professional establishment, capable of gathering intelligence, formulating policy, planning and directing operations, and recruiting, training, equipping, transporting and administering a modern armed force."

The "machine age" gave birth to the machine gun, to mechanized warfare, and to entirely new kinds of firepower, which, in turn, led inevitably, as we shall see, to new kinds of tactics. Industrialization led to improved roads, harbors, energy supplies, and communication. It gave the modern nation-state more efficient means of tax collection. All these developments vastly enlarged the scale of potential military operations.

As the Second Wave surged through society, First Wave institutions were eroded and washed away. A social system arose that linked mass production, mass education, mass communication, mass consumption, mass entertainment with, increasingly, weapons of mass destruction.

DEATH ON THE ASSEMBLY LINE

Relying on its industrial base for victory, the United States during World War II not only sent 15 million men to war, but mass-manufactured nearly 6 million rifles and machine guns, over 300,000 planes, 100,000 tanks and armored vehicles, 71,000 naval vessels, and 41 billion (*billion*, not million) rounds of ammunition.

World War II showed the awesome potential for industrializing death. The Nazis murdered 6 million Jews in true factory style —

creating what were, in effect, assembly lines for death. The war itself led to the slaughter of 15 million soldiers from all countries and nearly twice that number of civilians.

Thus, even before atomic bombs destroyed Hiroshima and Nagasaki, war had reached unparalleled levels of mass destruction. On March 9, 1945, for example, 334 American B-29 bombers hit Tokyo in a single attack that destroyed 267,171 buildings and killed 84,000 civilians (wounding 40,000 more), while flattening 16 square miles of the city.

Massive raids also hit Coventry, in England, and Dresden, in Germany, not to mention smaller population centers all across Europe.

Unlike Sun-tzu, who held that the most successful general was the one who achieved his ends without battle, or with minimal losses, Karl von Clausewitz (1780–1831), the father of modern strategy, taught a different lesson. While in later writings he made many subtle and even contradictory points, his dictum that "war is an act of violence pushed to its utmost bounds" reverberated through the wars of the industrial age.

BEYOND THE ABSOLUTE

Clausewitz wrote of "absolute war." That, however, was not enough for some of the theorists who followed. Thus the German general Erich Ludendorff after World War I expanded the concept to "total war," in which he stood Clausewitz on his head. Clausewitz saw war as an extension of politics, and the military as an instrument of political policy. Ludendorff argued that for war to be total the political order itself had to be subordinated to the military. Nazi theorists later extended even Ludendorff's notions of total war by denying the reality of peace itself and insisting that peace was merely a period of war preparation — "the war between wars."

In its larger sense, total war was to be waged politically, economically, culturally, and propagandistically, and the entire society converted into a single "war machine." It was industrial-style rationalization carried to its ultimate.

The military implication of such theories was maximization of destruction. As B. H. Liddell Hart wrote in his history of strategic thinking, "For more than a century the prime canon of military doctrine has been that 'the destruction of the enemy's main forces on the battlefield' constituted the only true aim in war. That was universally accepted, engraved in all military manuals, and taught in all staff colleges. . . . So absolute a rule would have astonished the great commanders and teachers of war-theory in ages prior to the nineteenth century."

But those ages were still largely preindustrial. The concepts of total war and mass destruction were widely adopted after the industrial revolution because they fit the ethos of a mass society — the civilization of the Second Wave.

In practice, total war blurred or completely eliminated the distinction between military and civilian targets. Since everything supposedly contributed to a total war effort, everything — from arms warehouses to workers' housing, from munitions dumps to printing plants — was a legitimate target.

Curtis LeMay, the general who led the Tokyo raid and later became chief of the U.S. Strategic Air Command, was the perfect apostle of the theory of mass destruction. If war came, he insisted, there was no time for prioritization of targets, nor the technology for precise targeting. "To LeMay," writes Fred Kaplan in *The Wizards of Armageddon*, "demolishing everything was how you win a war . . . the whole point of strategic bombing was to be massive." LeMay was the keeper of America's nuclear bombers.

By the 1960s, with Soviet and NATO forces facing each other in Germany, "small" battlefield nuclear weapons were added to the arsenals of the superpowers. War scenarios pictured the use of these weapons and the deployment of "vast tank formations" rolling forward over "a nuclear and chemical carpet" in the ultimate war of attrition.

Indeed, throughout the entire Cold War following World War II, the ultimate in mass destructive power, nuclear arms, dominated the relationship between the two great superpowers.

A DEADLY DOPPELGÄNGER

As industrial civilization reached its peak in the post–World War II period, mass destruction came to play the same central role in military doctrine as mass production did in economics. It was the deadly doppelgänger of mass production.

By the late 1970s and early 1980s, however, as Third Wave technologies, ideas, and social forms and forces began to challenge Second Wave mass society, a fresh breeze began to blow. It was, as we've seen, becoming clear to a small group of thinkers in the U.S. military and in the Congress that something was fundamentally wrong with American military doctrine. In the race to extend the range, speed, and lethality of weapons, the outer limits had already, for all practical purposes, been reached. The struggle against Soviet power had led to a nuclear standoff and insane threats of "mutually assured destruction." Was there a way to defeat Soviet aggression without nukes?

The development of modern war — the war of the industrial age — had reached its ultimate contradiction. A true revolution in military thinking was needed, a revolution that reflected the new economic and technological forces released by the Third Wave of change.

7

AIRLAND BATTLE

DONN STARRY is a tall, husky man, gray-haired and gray-eyed, who wears steel-rimmed glasses and speaks with quiet authority. He enjoys carpentry and painting his summer house in the solitary mountains of Colorado. He meticulously catalogs his 4,000-volume library. Once a year he and Letty, his wife, head for Canada, where they attend the Stratford Shakespeare Festival. He looks like a university president — which, in fact, he was for a time — although not at a conventional university.

Starry led the intellectual exercise that helped lift the U.S. Army from the black hole of demoralization into which it dropped after the Vietnam War to its peak performance in the Gulf War. He helped successfully restructure one of the biggest, most bureaucratic and recalcitrant institutions in the world — a task that very few captains of industry, dealing with far less cumbersome and complex organizations, have been able to accomplish.

In fact, largely unknown to the world, Starry's shadow hung over Saddam Hussein, the Iraqi dictator, all during the Persian Gulf War. For it was Donn Starry and Don Morelli who, as we saw earlier, began thinking about Third Wave warfare a decade before that war began.

Starry was a child of the Great Depression of the 1930s. His father worked in a furniture store for a while and for a local newspaper in the hard-hit farm country of Kansas. But he was also an officer in the Kansas National Guard, and Donn became the mascot of the weekend warriors in his hometown.

By 1943 the flames of World War II were spreading around the planet and Donn enlisted in the U.S. Army, eager to fight. But a perceptive first sergeant almost immediately tagged him as officer material. He led Starry to a batch of books he had selected and told him to lock himself in a room for three weeks and read those books. "Starry," the sergeant said, "you're going to take the competitive exam for West Point."

When Starry protested that he wanted to go to the front, his sergeant said, "Let me tell you something. This war isn't going to last forever. I've been in this army since World War I and the army will always need good officers. You wouldn't make one now — you're a lousy private. But I want you to go up there and study."

By the time Starry graduated from the army military academy as a second lieutenant, it was 1948. The war was over, and he was a young officer in a demobilizing army.

Starry rose in the ranks, up the normal ladder, from platoon leader and company commander to battalion staff officer. An expert on armor, he served in Korea in the 1950s as an intelligence officer on the Eighth Army staff. When U.S. involvement in the Vietnam War expanded in the 1960s, Starry served as a member of an army team analyzing mechanized and armored units and their functions.

Later, as a colonel, he commanded the famous Eleventh Armored Cavalry regiment during the U.S. incursion into Cambodia in 1970. There, in a skirmish near the airstrip at Snuol, he was wounded by a North Vietnamese grenade.

The American disaster in Vietnam, and especially the public derision heaped on the returning U.S. forces by an angrily divided country, embittered many officers and veterans. The military was attacked for drug use, corruption, atrocities. Men who had fought heroically found themselves accused of being "baby-killers." How could the most technologically advanced military in the world, one that actually won many conventional engagements with the North Vietnamese, be so ignominiously beaten by poorly clothed and equipped fighters from a Third World Communist nation?

THE JUNGLE TRAUMA

Like General Motors or IBM, the American military was almost perfectly organized for a Second Wave world. Like these corporations, it was designed for concentrated, mass, linear operations run from the top down. (Indeed, the war in Vietnam was micromanaged from the

White House itself, with the President sometimes personally selecting bomber targets.) It was heavily bureaucratic, torn by turf wars and branch rivalries. It did well when the North Vietnamese launched large-scale Second Wave operations. But it was poorly organized for small-scale guerrilla warfare — essentially First Wave warfare in the jungles.

What Starry calls "the army's miserable experience in Vietnam," however, had one positive effect. It led to a soul-searing self-analysis far deeper and more honest than that in most large corporations. The Vietnam trauma, according to Starry, "was so deeply embedded in everybody's minds that to do something new and different was very acceptable."

The crisis was even worse if one also looked at the military balance in Europe. While America was tied up in Vietnam, the Soviets had used the decade to modernize their tanks and missiles, to improve their doctrine, and to beef up their manpower in Europe. If the U.S. forces couldn't beat the North Vietnamese, what were their chances against the Soviet Red Army?

The Cold War was still the dominating fact of international life. While the United States had just suffered a humiliating defeat, the Soviet Union showed no signs yet of its future disintegration. Leonid Brezhnev and the Communist Party were still in power in Moscow. The Soviet military remained a seven-hundred-pound gorilla on the loose.

BOTTLING THE GENIE

Because Soviet and East Bloc conventional armies were so large, because their tanks so heavily outnumbered those of the West, NATO planners could see no way that their much smaller forces could beat back a Red Army attack on Western Europe without recourse to nuclear weapons. Indeed, virtually all NATO scenarios for the defense of Germany envisioned the use of nuclear weapons as early as three to ten days after the initial Soviet attack. But if nukes were used, they would destroy much of the West Germany NATO was pledged to defend.

Moreover, the ever-present threat of escalation from short-range tactical nukes to an all-out global nuclear exchange kept the lights blazing through the night at the Pentagon, at NATO headquarters in Brussels, and in the Kremlin as well.

That was the profound dilemma that Donn Starry faced when, in

1976, he was sent to command the U.S. Fifth Army Corps in Germany, posted at the most vulnerable spot in all of Europe. Here, at the Fulda Gap, near the city of Kassel, was the place where the Soviets were likely to attack first, if and when war broke out. If nuclear war began, it could well start here. In short, Starry suddenly found himself the West's point man against massive Soviet power.

For Starry, the central problem was clear: nobody must unleash the uncontrollable nuclear genie from its bottle. Therefore, the West must find a way to defend itself against the Soviet's overwhelming numerical superiority — without using its nuclear weapons. By the time he arrived to take command in Germany, Starry was convinced nonnuclear victory was possible. But not by reliance on the traditional doctrine.

A TICKET TO TEL AVIV

What convinced Starry was a short, savage conflict that had been waged three years earlier. For 2,000 miles to the east of the West German border, on the line between Israel and Syria, in the scraggy hills called the Golan Heights, one of the great tank battles in history had taken place. Tank officers everywhere would study this battle for decades to come.

It began on Yom Kippur Day, October 6, 1973, when, suddenly, the armies of Egypt and Syria attacked Israel. While the Israelis had made short work of the Arabs in the Six Day War in 1967, wiping out their air forces on the ground before they could climb into the sky, by 1973 the Arab forces were better equipped, better trained, and confident that once and for all they could defeat the Israelis. And why not?

The Syrian-led forces attacked in the north. Five divisions, with over 45,000 troops, backed by more than 1,400 tanks and 1,000 mortars and artillery pieces, hurled themselves across the Israeli border. The force included T62s, the most advanced Soviet tanks then made.

Facing them were two weak Israeli brigades, the Seventh in the northern sector and the 188th to the south — 6,000 men in all, with only 170 tanks and 60 pieces of artillery. Despite this glaring disparity, it was the Israelis, not the Syrians, who triumphed.

Two and a half months later, in early January 1974, Starry and a team of armor officers were invited by the British to visit some of their training facilities. Starry's wife, Letty, was with him. They were enjoying their off-hours together in England when suddenly a call came from Gen. Creighton Abrams, the army chief of staff. "There will be an officer on your doorstep tomorrow morning with all the necessary papers. Send your wife and staff home. Take one man with you. You're going to Israel."

Having spent the better part of his life studying tank warfare, Starry was determined to find out exactly what happened on the Golan Heights.

Soon Starry found himself gazing at the endless lines of destroyed Syrian tanks and burned-out personnel carriers. He walked every inch of the Golan battlefield. He met repeatedly with all the key Is-raeli commanders, Moshe "Mussa" Peled, Avigdor Kahalani, Benny Peled, and others at the battalion level, reliving every second of the battle.

SURPRISE AT KUNEITRA

The war had begun at 1:58 in the afternoon on October 6. Within twenty-four hours, the men of the 188th Brigade, attacked in the southern sector by two Syrian divisions with 600 tanks, had been wiped out. Ninety percent of their officers were dead or wounded, and the onrushing Syrians were within ten minutes of the Jordan River and the Sea of Galilee. The defenders seemed crushed, and the Syrians had almost overrun the Israeli divisional headquarters.

Meanwhile, the 500-tank Syrian force in the northern half of the Golan Heights struck with equal power at the Israeli Seventh Brigade defending with 100 tanks. There the battle raged for four days, during which the Seventh managed to destroy literally hundreds of the Syrian tanks and armored vehicles before its own tank force was reduced to seven. At that moment, short of ammunition and on the point of retreat, it was joined by thirteen additional tanks that had been damaged, hastily repaired, and sent back to fight, manned, in part by wounded men who discharged themselves from hospitals to return to battle. The Seventh Brigade, in one of the most heroic battles in Israeli history, launched a desperate surprise coun-terattack, at which point, to the Israelis' surprise, the exhausted Syrians withdrew.

The audacious, seemingly hopeless struggle of the Seventh Brigade in the northern sector is now memorialized in a firsthand account called *The Heights of Courage*, written by Avigdor Kahalani, a battal-ion commander in the Seventh, and bearing a preface by Donn Starry.

But the really key battle took place in the southern sector. And it was this engagement that changed the way Starry thought about war.

The bloody stand of the Seventh Brigade in the north gained just enough time for reinforcements to arrive in the south. One division, commanded by Gen. Dan Laner, approached from the southwest. A second, under Gen. Moshe "Mussa" Peled, made a parallel approach about ten miles to the south of Laner's force. These forces, now with intense support from the Israeli air force, closed toward one another to form a pincer around a concentration of Syrian forces a few miles south of Kuneitra.

Starry closely questioned the Israeli commanders about every detail of that battle. At one point, he learned, an argument had broken out among them about what to do with the reinforcements under "Mussa" Peled. They were supposed to strengthen the weakest points and continue to defend. But Peled objected. All that would do, he contended, was lead to further attrition — and eventually to defeat. Instead, Peled — supported by General Chaim Bar-lev, a former chief of staff, who was then a top military adviser to Prime Minister Golda Meir — decided to use his reinforcements to attack. In the midst of general defeat, a tactical attack was ordered, and, instead of directing it at the main point of Syrian strength, it would strike at them from an unexpected direction.

Even though Peled lost many men, his attack on the left of the Syrian forces surprised and threw them off balance. With Laner's advance the pincer closed on them. The result was not just a surprise, but a rout. It meant that many of the Syrian backup forces could not come into play.

"By midday on Wednesday 10 October," writes Chaim Herzog in *The Arab-Israeli Wars*, "almost exactly four days after some 1,400 Syrian tanks had stormed across the Purple Line* in a massive attack against Israel, not a single Syrian tank remained in fighting condition West of that line."

Soon the Israelis regrouped and pushed into Syria itself, almost to its capital, Damascus. Behind them, writes Herzog, "the pride of the Syrian army lay smoking and burnt out along their earlier axes of advance. . . . The most modern arms and equipment that the Soviet Union had supplied to any foreign army dotted the undulating hills of the Golan heights, testimony to one of the great tank victories in history against almost incredible odds."

By the time a UN cease-fire was accepted by the Syrians, ending the war, they had lost 1,300 tanks (of which 867 fell into the hands of

---

* The cease-fire line separating Syria from Israel after the Six Day War in 1967.

the Israelis). Some 3,500 Syrians had died and another 370 had been captured. All Israeli tanks had been hit at one time or another, but many had been instantly repaired and thrown back into battle. Only about 100 were totally destroyed. The Israelis lost 772 men, and another 65 were imprisoned by the Syrians.

The primary lesson, for Starry, was that "starting ratios" do not determine the outcome. "It makes no difference who is outnumbered or who is outnumbering." Put differently, the fact that the Syrians had echelon after echelon of backup troops did them no good at all.

The other unmistakable lesson was that whoever seizes the initiative, "whether he is outnumbered or outnumbering, whether he is attacking or defending," will win. As the Israelis showed, even a small army strategically on the defensive might be able to seize the initiative.

These ideas were not new. But they flew directly in the face of then-conventional thought. The old assumption — one embedded in war games and training maneuvers — was that if the Soviets ever attacked in Germany, NATO troops would retreat, fight a delaying action, then go over to the offense and push them back. If they failed, they would fall back on nuclear weapons.

That, Starry concluded, was wrong. "I realized that we had to delay and disrupt, deep into the enemy's battle area. The orderly advance of their follow-on echelons would have to be stopped. We wouldn't have to destroy them. It would be nice if we could. But all we really had to do was prevent them from getting to the battle, so they couldn't overwhelm the defenders."

ACTIVE DEFENSE

If masses of Soviet-supplied Syrians, using Soviet doctrine, could be stopped by heavily outnumbered Israelis making a shallow encirclement, Starry reasoned, why couldn't masses of Soviet and Eastern European troops also be stopped by smaller allied forces — without the use of nuclear weapons? In fact, the lessons might be applicable to other parts of the world as well, where various countries were building huge conventional armies based on the old doctrine that sheer mass wins.

Persuaded by the Vietnam disaster that change was desperately needed, the U.S. Army in 1973 had created TRADOC — the Training and Doctrine Command, under Gen. William E. DePuy. Hardly known to the public, TRADOC runs the largest educational system in the non-Communist world. It operates the equivalent of many universities for officers, along with literally hundreds of training centers. It devotes great attention to things like learning theory and advanced training technologies. But it also provides much of the theoretical underpinning for the army's conception of warfare. And inside TRADOC, within a year or two of its founding, a post-Vietnam intellectual ferment began brewing.

In 1976, about the time Starry was posted to Germany, TRADOC issued a new army doctrine entitled Active Defense. Drawing in part on the Israeli experience and input from Starry, it argued for "deepening" the battleground — striking not merely at the first echelon of any invading Soviet force, but using high-tech weapons with longer range to take out the next echelon of backup troops as well.

This doctrine was a step in the right direction as far as Starry was concerned. But the second echelon of an advancing Red Army was not the only problem. What about the third, the fourth, and the echelons after that? There were a lot more Soviet troops than Syrians. Active Defense did not go nearly far enough in rethinking warfare.

CHANGING THE PENTAGON

The need for deeper reconceptualization was still haunting Starry when, in 1977, he himself was promoted and sent to take over TRADOC.

Starry is always careful to credit Active Defense doctrine and General DePuy, with whose views he now says he agreed almost entirely. But at the time there was a strong difference between them on the issue of defense versus offense. What was needed, Starry concluded, was not just an incremental change, but a total rethink of the U.S. Army's doctrine from the ground up.

Moreover, while the debate over these issues was under way inside the military, American society, in which the military was embedded, was itself undergoing deep change. New ideas and new possibilities were in the air. Thus as the American economy began moving decisively away from old-style mass production toward de-massified production, as a Third Wave system for creating wealth began to take form, the U.S. Army began a parallel development. Though the outside world remained unaware of it, the first steps were being taken to formulate a theory of Third Wave war.

Starry's attempt to force that "rethink" made him challenge some of the key assumptions of Second Wave warfare. It forced him into the role of doctrinal revolutionary, triggering a process that is still unfolding and taking new directions.

Changing any military's doctrine, however, is like trying to stop a tank armor by throwing marshmallows at it. The military, like any huge modern bureaucracy, resists innovation — especially if the change implies the downgrading of certain units and the need to learn new skills and to transcend service rivalries.

To define a new doctrine, to win support for it both in the armed forces and among politicians, and then to actually implement it with trained troops and appropriate technologies is a tremendous task, and no one man, general or not, could possibly hope to accomplish it. It would take a campaign — one in which ideas would be the bullets.

The campaign began with military intellectuals, spurred by Starry, writing papers and publishing them in the military equivalent of scholarly journals. Reviewers — the military version of literary critics — tore the various papers and proposals apart in a lengthy, complex intellectual process.

Key to this effort was a reexamination of the old obsession with sheer mass. To question that meant challenging not simply an idea but all the jobs, careers, tactics, technologies, and industrial relationships based on it. It meant reviewing and possibly changing the entire force structure of the army — that is, the size, composition, and number of units in it. And it meant doing this at a time when the formal Soviet doctrine was still actually named "Mass Momentum and Continuous Land Combat." Indeed, questioning the idea of mass not only flew in the face of military doctrine, it ran counter to the ethos of industrial mass society.

The breakthrough to a new concept of warfare only crystallized in the late 1970s and early 1980s. In this period, Starry read widely, not just about military matters but about the new social and economic forces moving us beyond modernity, from a Second Wave toward a Third Wave civilization. It was in the course of this study that he read our book *The Third Wave* and recommended it to the generals on his staff.

"The army," he told us in 1982 at our first meeting, "is very hard to change. After all, it is a . . . Second Wave institution. It's a factory. The idea was that our industrial factories will produce and produce and produce weapons. The army will run men through a training factory. Then it will bring the men and the weapons together and we'll win wars. The entire approach is Second Wave. It needs to be brought into the Third Wave world."

To carry out this mission, Starry needed the support of his superiors. He got it from Gen. E. C. Meyer, then the army chief of staff, from his predecessor at TRADOC, Bill DePuy, from General Abrams, and others. These men assured Starry that disagreement would not be regarded as disloyalty. Still wracked by the Vietnam trauma, they, too, understood that fresh thinking was essential.

Starry also needed extremely sophisticated officers — military intellectuals — on his staff. And he proceeded to bring them to TRADOC headquarters in Fort Monroe, Virginia. In addition, Gen. William R. Richardson and a small flock of colonels —Richmond Henriques, Huba Wass de Czege, and L. D. Holder —worked for Starry at Fort Leavenworth in Kansas, helping to define the problems and work out the implications of any doctrinal change.

Starry also took steps to upgrade the development of doctrine, often in the past relegated to secondary status. He did this by creating the new post of Deputy Chief of Staff for Doctrine. One day Don Morelli walked into his office. And in short order Brigadier General Morelli was placed in charge of the new office of doctrine formulation.

Starry and Morelli, and a small group of other officers — James Merryman, Jack Woodmansee, Carl Vuono, along with a civilian, Dr. Joe Braddock (whose consulting firm, Braddock, Dunn and MacDonald, or BDM, worked for the Defense Nuclear Agency) formed a floating think tank for TRADOC.

As they hammered out their ideas about weapons, organization, logistics, electronic warfare, the threat of nuclear weapons, and the importance of maneuver as against positional warfare, Starry and Morelli traveled incessantly, trying out their concepts in briefings of military audiences all over the United States, Britain, and Germany. Questions and criticisms sharpened their minds.

Meanwhile, at home, there were interservice problems. The air force had no exact counterpart of TRADOC. The closest equivalent then was TAC, the Tactical Air Command, at Langley Air Force Base, just fifteen minutes away from Fort Monroe (one of the reasons TRADOC was placed there).

Starry's emphasis on the concept of "deep battle," or the "extended battlefield," meant that combat would not simply take place at the "front" but deep in the enemy's rear as well — back where the follow-on echelons were to be found. It was necessary to "interdict" the movement of men, supplies, and information so that the rear echelons would not be able to support invading troops.

Deep strikes by the air force would be needed to knock out the adversary's command centers, logistic lines, communication links, and air defenses. This, in turn, would require the closest integration of air and ground forces. But there were elements in the air force who regarded all such discussion with suspicion. It seemed to them (and to some air force officers even today) that the army was attacking air force turf, trying to engage in interdiction, traditionally an air force responsibility.

It was the commander of TAC, Bill Creech, who persuaded his superiors that the development of doctrine for a new way of fighting wasn't a matter of turf. Soon a team of air force officers was working side by side with the TRADOC men on a daily basis, trying to hammer out the appropriate relationships of air and ground activities.

Even while developing the doctrine, Starry had to answer questions about implementation. What kind of soldiers and officers would be needed in the future? And what technologies would they need?

TRADOC was charged not only with formulating a new doctrine and training a new-style army but with actually determining what types of weapons and technologies that army of the future would need. Thus TRADOC, in fact, helped define the requirements for M-1 Abrams tanks, Apache helicopters, the Bradley fighting vehicle, and the Patriot missile — weapons not yet out of production at the time. J-STARS, the widely praised air-based radar system that provided detailed targeting information to ground stations during Desert Storm, was similarly hatched in TRADOC in 1978–79. The MRLS, or multiple rocket launcher system, the ATACMS missile system, all were among the weapons that TRADOC determined, years in advance, would be necessary to implement its new fighting doctrine.

Out of this intense activity, at last, on March 25, 1981, came the first formal statement of the new future-focused doctrine. It was a thin Xeroxed pamphlet in a camouflage-green cover entitled *The Air-Land Battle and Corps 86, TRADOC Pamphlet 525-5*. This was a preliminary paper that Morelli (who coined the term AirLand Battle) used in his busy schedule of briefings, now reaching outside the military to members of Congress, White House officials, the vice president, and even — as earlier noted — to us, a pair of decidedly nonmilitary intellectuals.

The concept of AirLand Battle was now out in the open — subject to outside analysis, attack, and criticism not only from politicians and traditionalists in the U.S. military, but from many in the NATO nations in Europe who saw in it not a way to avoid nuclear war but merely evidence of America's "aggressive" spirit.

The Starry-Morelli doctrine was finally embodied in the army's *Field Manual (FM) 100-5 (Operations)* on August 20, 1982, some four months after our first contact with Morelli. It would become, as he wished, the basis for similar or parallel doctrinal changes in Western European armies in NATO. It emphasized close air and land coordination, deep strikes to prevent first, second, and subsequent echelons from reaching the scene of battle, and — most significantly —the use of new technologies to hit targets previously assigned to nuclear weapons. In doing so, it reduced the chances of nuclear confrontation.

Emphasizing the lesson Starry brought back from the Golan Heights, the new manual urged officers and men to seize the initiative — to go on the offensive tactically or operationally, even when on the defensive strategically. Even if a powerful enemy has broken through, as the Syrians did at first, surprise counterattacks should be aimed at its weak spots, rather than frontally against the decisive point of breakthrough. Finally, the new doctrine hammered away at the need for higher human quality — not only leadership but training to increase each soldier's capabilities.

Since it first appeared, AirLand Battle doctrine has been updated, refined, and renamed. Whereas AirLand Battle aimed at disrupting an enemy's rear echelons, a later version entitled AirLand Operations urges early action to prevent the rear echelons from forming in the first place. Work on AirLand Operations began in 1987. It became official doctrine on August 1, 1991 — one year after Saddam Hussein surprised the world by invading Kuwait.

It emphasized the capacity to project power long distances at high speed. It stressed the need for joint operations among the different services and combined operations with allied forces. It called for "greater scope for initiative" and "greater reliance on quality soldiers."

Placing time at the center of its concerns, it called for synchronized simultaneous attacks and "execution control in real time." Commanders should "control the tempo of flights." Finally, knowledge — improved intelligence and communication — becomes absolutely central.

So accelerated are changes in the world scene these days that doctrinal revisions — which used to take place at forty- or fifty-year intervals — now are needed every year or two.

Thus on June 14, 1993, the latest revision of the *Field Manual*

(FM) 100-5 appeared. "Recent experiences gave us a glimpse of new methods of warfare," declares the executive summary of the newest doctrine. "They were the end of industrial-age warfare and the beginning of warfare in the information age."

This newest version places high stress on versatility — the ability of the army to switch from one kind of conflict to another quickly. It shifts from a European to a global focus and from the idea of forward deployment — that is, forces based near zones of potential conflict — to the idea of a U.S.-based force that can go anywhere in the world fast. It moves from a preoccupation with the threat of global war with the Soviets to an emphasis on regional contingencies. In addition, the new doctrine devotes attention to what it calls "operations other than war," which, in its terms, include disaster relief, civil disturbance, peacekeeping, and counter-narcotics activities.

It explains carefully that the U.S. Army is responsible to the American people who "expect quick victory and abhor unnecessary casualties" and who "reserve the right to reconsider their support should any of these conditions not be met."

The latest revision is thoughtful and timely. (As an intellectual product it merits attention in the *New York Times Book Review*.) It reflects some of the dramatic changes in the global situation since AirLand Battle was written and thus reaches far beyond AirLand Battle. Nevertheless, as in the case of the earlier revisions, its DNA is still to be found in the Starry-Morelli doctrine, the U.S. military's first conscious attempt to adapt to the Third Wave of change.

To understand all that follows, we need to look at the impact of this work in a war that uncannily mirrored the rise of a new form of economy — the revolutionary Third Wave system for wealth creation.

8

THE WAY
WE MAKE WEALTH...

IN 1956 the Soviet Union's roly-poly strongman Nikita Khrushchev uttered his famous boast — "We will bury you." What he meant was that communism would outstrip capitalism economically in the years ahead. The boast carried with it, as well, the threat of military defeat, and it reverberated around the world.

Yet few at the time even dimly suspected just how a revolution in the West's system for creating wealth would transform the world military balance — and the nature of warfare itself.

What Khrushchev (and most Americans) didn't know was that 1956 was also the first year in which white-collar and service employees outnumbered blue-collar factory workers in the United States — an early indication that the Second Wave's smokestack economy was fading and a new, Third Wave economy was being born.

Before long a few futurists and pioneer economists began tracking the growth of knowledge-intensivity in the U.S. economy and trying to anticipate its long-term impact. As early as 1961 IBM asked a consultant to prepare a report on the long-term social and organizational implications of white-collar automation (many of its conclusions still valid today). In 1962 economist Fritz Machlup published his groundbreaking study, *The Production and Distribution of Knowledge in the United States.*

In 1968, AT&T, then the world's largest private corporation, commissioned a study to help it redefine its mission. In 1972, a decade before it was dismantled by the U.S. government, it received that report — a heretical document urging the firm to restructure itself drastically and to break itself up.

The report outlined the ways in which a giant Second Wave, industrial-style bureaucracy might transform itself into a fast-moving, maneuverable organization. But AT&T suppressed the report for three years before allowing it to circulate in top management. Most major American companies had not yet begun to think beyond incremental reorganization. The notion that radical surgery would be needed for them to survive in the emergent knowledge-based economy seemed exaggerated. Yet the Third Wave soon hurled many of the world's biggest organizations into the most painful restructuring in their history.

Thus in the same rough time frame in which Starry and his supporters were beginning to reshape U.S. military thinking, many of America's giant companies also began to cast about, looking for new missions and new organizational structures. A flurry of new management doctrines arose as the very method of creating wealth changed.

To understand the extraordinary changes in warfare that have since occurred, and to anticipate the even more dramatic changes that lie ahead, we need to look at ten key features of the new Third Wave economy.

1. FACTORS OF PRODUCTION

While land, labor, raw materials, and capital were the main "factors of production" in the Second Wave economy of the past, knowledge — broadly defined here to include data, information, images, symbols, culture, ideology, and values — is the central resource of the Third Wave economy. Once scoffed at, this idea has already become a truism. Its implications, however, are still little understood.

Given the appropriate data, information, and/or knowledge, it is possible to reduce all the other inputs used to create wealth. The right knowledge inputs can reduce labor requirements, cut inventory, save energy, save raw materials, and reduce the time, space, and money needed for production.

A computer-driven cutting tool, operating with exquisite precision, wastes less cloth or steel than the pre-intelligent cutting machine it replaces. "Smart" automated presses that print and bind books use less paper than the brute-force machines they replace. Intelligent controls save energy by regulating the heat in office buildings. Electronic data systems linking manufacturers to their customers reduce the amount of goods — from capacitors to cotton wear — that must be kept in inventory.

Thus knowledge, used properly, becomes the ultimate substitute for other inputs. Conventional economists and accountants still have trouble with this idea, because it is hard to quantify, but knowledge is now the most versatile and the most important of all the factors of production, whether it can be measured or not.

What makes the Third Wave economy truly revolutionary is the fact that while land, labor, raw materials, and perhaps even capital can be regarded as finite resources, knowledge is, for all intents, inexhaustible. Unlike a single blast furnace or assembly line, knowledge can be used by two companies at the same time. And they can use it to generate still more knowledge.

2. INTANGIBLE VALUES

While the value of a Second Wave company might be measured in terms of its hard assets like buildings, machines, stocks, and inventory, the value of successful Third Wave firms increasingly lies in their capacity for acquiring, generating, distributing, and applying knowledge strategically and operationally.

The real value of companies like Compaq or Kodak, Hitachi or Siemens, depends more on the ideas, insights, and information in the heads of their employees and in the data banks and patents these companies control than on the trucks, assembly lines, and other physical assets they may have. Thus capital itself is now increasingly based on intangibles.

3. DE-MASSIFICATION

Mass production, the defining characteristic of the Second Wave economy, becomes increasingly obsolete, as firms install information-intensive, often robotized manufacturing systems capable of endless, cheap variation, even customization. The revolutionary result is, in effect, the de-massification of mass production.

The shift toward smart "flex-techs" promotes diversity and feeds consumer choice to the point that a Wal-Mart store can offer the buyer nearly 110,000 products in various types, sizes, models, and colors to choose among.

But Wal-Mart is a mass merchandiser. Increasingly, the mass market itself is breaking up into differentiated niches as customer needs diverge and better information makes it possible for businesses to identify and serve micro-markets. Specialty stores, boutiques, superstores, TV home-shopping systems, computer-based buying, direct mail, and other systems provide a growing diversity of channels through which producers can distribute their wares to customers in an increasingly de-massified marketplace.

Meanwhile, advertising is targeted at smaller and smaller market segments reached through increasingly de-massified media. The dramatic breakup of mass audiences is underscored by the crisis of the once great TV networks, ABC, CBS, and NBC, at a time when Tele-Communications Inc. of Denver announces a fiber-optic network capable of providing viewers with 500 interactive channels of television. Such systems mean that sellers will be able to target buyers with even greater precision. The simultaneous de-massification of production, distribution, and communication revolutionizes the economy, and shifts it from homogeneity toward extreme heterogeneity.

4. WORK

Work itself is transformed. Low-skilled, essentially interchangeable muscle work drove the Second Wave. Mass, factory-style education prepared workers for routine, repetitive labor. By contrast, the Third Wave is accompanied by a growing non-interchangeability of labor as skill requirements skyrocket.

Muscle power is essentially fungible. Thus a low-skilled worker who quits or is fired can be replaced quickly and with little cost. By contrast, the rising levels of specialized skills required in the Third Wave economy make finding the right person with the right skills harder and more costly.

Although he or she may face competition from many other jobless muscle workers, a janitor laid off from a giant defense firm can take a janitor's job in a school or an insurance office. By contrast, the electronics engineer who has spent years building satellites does not necessarily have the skills needed by a firm doing environmental engineering. A gynecologist can't do brain surgery. Rising specialization and rapid changes in skill requirements reduce the interchangeability of labor.

As economies advance, a further change is seen in the ratio of "direct labor" to "indirect labor." In traditional terms (fast losing their significance) direct, or "productive," workers are those on the factory floor who actually make the product. They produce added value, and everyone else is described as "nonproductive" or making only an "indirect" contribution.

Today these distinctions blur as the ratio of factory production workers to white-collar, technical, and professional workers declines, even on the factory floor. At least as much value is produced by "indirect" as by "direct" labor — if not more.

5. INNOVATION

With the economies of Japan and Europe recovered from World War II, American firms face heavy competitive fire. Constant innovation is needed to compete — new ideas for products, technologies, processes, marketing, finance. Something on the order of 1,000 new products are introduced into America's supermarkets every month. Even before the model 486 computer has replaced the model 386 computer, the new 586 chip is on its way. Thus smart firms encourage workers to take initiative, come up with new ideas, and even, if necessary, to "throw away the rule book."

6. SCALE

Work units shrink. Rather than thousands of workers pouring into the same factory gate — the classic image of the smokestack economy — the scale of operations is miniaturized along with many of the products. The vast numbers of workers doing much the same muscle work are replaced by small, differentiated work teams. Big businesses are getting smaller; small businesses are multiplying. IBM, with 370,000 employees, is being pecked to death by small manufacturers around the world. To survive it lays off many workers and splits itself into thirteen different — smaller — business units.

In the Third Wave system, economies of scale are frequently outweighed by diseconomies of complexity. The more complicated the firm, the more the left hand can't anticipate what the right hand will do next. Things fall through the cracks. Problems proliferate that may outweigh any of the presumed benefits of sheer mass. The old idea that bigger is necessarily better is increasingly outmoded.

7. ORGANIZATION

Struggling to adapt to high-speed changes, companies are racing to dismantle their bureaucratic Second Wave structures. Industrial-era companies typically had similar tables of organization — they were pyramidal, monolithic, and bureaucratic. Today's markets, technologies, and consumer needs change so rapidly, and put such varied pressures on the firm, that bureaucratic uniformity is on its way out. The search is on for wholly new forms of organization. "Re-engineering," for example, the current buzzword in management, seeks to restructure the firm around processes rather than markets or compartmentalized specialties.

Relatively standardized structures give way to matrix organizations, ad hocratic project teams, profit centers, as well as to a growing diversity of strategic alliances, joint ventures, and consortia — many of these crossing national boundaries. Since markets change constantly, position is less important than flexibility and maneuver.

8. SYSTEMS INTEGRATION

Rising complexity in the economy calls for more sophisticated integration and management. In a not atypical case, Nabisco, the food company, has to fill 500 orders a day for literally hundreds of thousands of different products that must be shipped from 49 factories and 13 distribution centers and, at the same time, take into account 30,000 different sales promotional deals with its customers.

Managing such complexity requires new forms of leadership and an extremely high order of systemic integration. That, in turn, requires higher and higher volumes of information to pulse through the organization.

9.

INFRASTRUCTURE

To hold everything together — to track all the components and products, to synchronize deliveries, to keep engineers and marketers apprised of each other's plans, to alert the R & D people to the needs of the manufacturing side, and, above all, to give management a coherent picture of what is going on — billions of dollars are being poured into electronic networks that link computers, data bases, and other information technologies together.

This vast electronic information structure, frequently satellite-based, knits whole companies together, often linking them into the computers and networks of suppliers and customers as well. Other networks link networks. Japan has targeted $250 billion to develop better, faster networks over the next twenty-five years. U.S. vice president Gore, when still in the Senate, sponsored legislation that provides $1 billion over five years to help start up a "National Research and Education Network" intended to do for information what superhighways did for cars. Such electronic pathways form the essential infrastructure of the Third Wave economy.

10. ACCELERATION

All these changes further accelerate the pace of operations and transactions. Economies of speed replace economies of scale. Competition is so intense and the speeds required so high, that the old "time is money" rule is increasingly updated to "every interval of time is worth more than the one before it."

Time becomes a critical variable as reflected in "just-in-time" deliveries and a pressure to reduce DIP or "decisions in process." Slow, sequential, step-by-step engineering is replaced by "simultaneous engineering." Companies wage "time-based competition." Expressing the new urgency, DuWayne Peterson, a top executive at Merrill Lynch, says, "Money moves at the speed of light. Information has to move faster." Thus acceleration pushes Third Wave business closer and closer toward real time.

Taken together, these ten features of the Third Wave economy, among many others, add up to a monumental change in how wealth is created. The conversion of the United States, Japan, and Europe to this new system, though not yet complete, represents the single most important change in the global economy since the spread of factories brought about by the industrial revolution.

This historical transformation, picking up speed in the early- to mid-seventies, was already fairly well advanced by the 1990s. During this period, war itself began to be transformed in tandem. Second Wave war, like Second Wave economics, was racing toward obsolescence.

9

THIRD WAVE WAR

SOMETHING occurred in the night skies and desert sands of the Middle East in 1991 that the world had not seen for three hundred years — the arrival of a new form of warfare that closely mirrors a new form of wealth creation. Once again, we find that the way we make wealth and the way we make war are inextricably connected.

The world's most technologically advanced societies today have split-level economies — partly based on declining Second Wave mass production, partly on emergent Third Wave technologies and services. None of the high-tech nations, not even Japan, has completed its transition to the new system of economics.

Even the most advanced economies — those in Europe, Japan, and the United States — are still divided between declining muscle work and increasing mind work. This duality was sharply reflected in the way the Gulf War of 1990–91 was fought.

However history may ultimately evaluate that conflict in terms of morality, economics, or geopolitics, the actual way in which the war was fought held — and still holds — profound implications for armies and for countries all over the world.

What is not clearly understood even now is that the United States and its allies simultaneously fought two very different wars against Iraq's Saddam Hussein. More accurately, it applied two different war-forms, one Second Wave, the other Third Wave. The Gulf bloodshed began on August 2, 1990, with Saddam Hussein's attack on neighboring Kuwait — not, as is so often said, on January 17, 1991, when the U.S.-led coalition struck back at Baghdad. Saddam drew first blood.

In the months that followed, as the United States and the United Nations coalition debated how to respond, Saddam boasted that the allies would find themselves ground to shreds in the "Mother of All Battles." His theme was picked up by Western media pundits and politicians who predicted huge allied losses, some as high as 30,000 killed. Even some military analysts concurred.*

TECHNOPHOBIA

Simultaneously, some opponents of the war launched what seemed like a campaign in the Western media against advanced technology itself. The world press soon echoed with technophobic rhetoric. U.S. helicopters would be downed by sandstorms. The stealth bomber would fail. Night-vision goggles wouldn't work. Dragon and TOW anti-tank weapons would be useless against "Soviet-supplied Iraqi armor." The M-1 tank would prove ineffective and break down frequently. "Is Our High-Tech Military a Mirage?" the New York Times wanted to know.

One leading military columnist dismissed the whole idea that technology could "tip the odds" in warfare. This, he informed his readers, was a "myth" and Americans were profoundly mistaken when they "emphasized materiel over manpower."

Some "military reformers" on Capitol Hill, voicing a familiar refrain, attacked advanced weaponry as "too complex to work." They argued, as they had for years, that what the United States needed was masses of simpler planes, tanks, and missiles, rather than smaller numbers of more sophisticated weapons.

All this added to the growing public dread of huge allied losses. After all, Saddam had a million-man, Soviet-indoctrinated, Soviet-supplied army. Unlike the allied forces, it was battle-tested and recently blooded in an eight-year war against Iran. Moreover, it had six months to dig in, build berms, bunkers, and trenches and to lay murderous minefields. The Iraqis, it was predicted, would set fire to oil-filled ditches and create an uncrossable line of flames. Supporting their first-line troops, the Iraqis had deployed echelon after backup echelon of massed men and armor (like the Syrians at the Golan Heights or the Soviets in Central Europe). If allied ground troops dared to attack, they would be decimated.

---

* Actual losses were approximately 340 — roughly one hundredth of these forecasts.

Saddam Hussein had only to wait for America to become politically demoralized by television images of vast numbers of body bags arriving at U.S. military cemeteries. Political resolve would collapse. And he could keep Kuwait, or at least its oil-rich regions.

This, however, presupposed that the war in the Gulf would be a typical industrial-era war. Though the basic ideas in AirLand Battle (and its later revisions) were already common currency in military circles around the world, Saddam, despite his pretensions to military expertise, seemed totally unaware of them. Saddam never understood that an entirely new war-form was about to change the entire nature of warfare.

The dual war began with the earliest allied attacks.

THE DUAL WAR

From the outset there were two air campaigns, although they were integrated and few thought of them as separate. One employed the familiar attrition-style methods of modern — that is, Second Wave — war. Fleets of thirty-year-old aircraft relentlessly carpet-bombed the Iraqis in their bunkers. Just as in previous wars, "stupid" bombs were dropped, causing widespread destruction casualties, creating havoc, and demoralizing both the Iraqi front-line troops and the backup Republican Guards. The coalition commander, General Schwarzkopf, was "preparing the battlefield," as his press briefers put it, while half a million allied ground troops stood poised to move against the Iraqi line.

In Paris after the war, the authors spoke with retired Gen. Pierre Gallois. Formerly in the French Air Force and later assistant to the commander of NATO, responsible for strategic studies, Gallois visited Iraq immediately after the fighting. "I drove for twenty-five hundred kilometers in my four-wheel-drive," he told us, "and in the villages, everything was destroyed. We found bomb fragments dated from 1968, left over from the Vietnam War. This was the same kind of bombing I did half a century ago in World War Two."

This most murderous form of warfare was well understood by both sides. It was industrialized slaughter, and we will never know how many Iraqi troops and civilians died as a result.

But a radically different kind of war was also waged from Day One. The world was stunned at the very start by unforgettable television images of Tomahawk missiles and laser-guided bombs searching out and hitting their targets in Baghdad with astonishing accuracy: the Iraqi Air Force headquarters, the buildings housing the Iraqi Intelligence Service, the Ministry of the Interior (headquarters of Saddam's police), the Congress Building, the headquarters of his Ba'ath Party.

Because of their ability to penetrate high-threat areas and to deliver precision-guided bombs, Nighthawk stealth fighters — otherwise known as F-117As — were the only planes to attack targets in downtown Baghdad. They focused on well-protected air-defense centers and military command and control facilities. Flying only 2 percent of total sorties, they accounted for 40 percent of strategic targets attacked. And, despite all the gloomy forecasts, every one returned safely.

Throughout the remaining days of conflict, television accentuated this new war-form. Missiles virtually went around corners and entered pretargeted windows in bunkers hiding Iraqi tanks and troops. War was seen on our TV screens as it appeared on the electronic monitors of the pilots and soldiers doing the fighting.

The result was a highly sanitized image of war, a seemingly bloodless form of combat in stark contrast to the TV coverage of the Vietnam War, which hurled dismembered limbs, shattered skulls, and napalmed babies into the American living room.

One war in Iraq was fought with Second Wave weapons designed to create mass destruction. Very little of that war was shown on the world's video screens; the other war was fought with Third Wave weapons designed for pinpoint accuracy, customized destruction, and minimal "collateral damage." That war was shown.

Many of the key weapons systems employed by the United States were built, as we saw, to meet requirements defined by Starry's TRADOC in the preceding decade. But the imprint of Starry, who was already retired by the time the war broke out, and of Morelli, dead for almost a decade, was even more strongly evident in the way the weapons were used.

For example, from the beginning of combat, it reflected their thinking about "deep battle," "interdiction," and the importance of information and intelligent weapons.

THE VANISHING FRONT

During World War I, millions of soldiers had faced each other from fortifications dug into the soil of France. Filled with mud and rats, stinking of garbage and gangrene, these linear trenches stretched for miles across the countryside, behind tangles of barbed wire. For months at a time whole armies crouched, afraid to raise a head above ground level. When an attack was ordered, the troops would go "over the top" and face a hurricane of artillery and small-arms fire. But for the most part they sat, immobilized as disease and ennui spread through the ranks.

There was little question in anyone's mind where the "front line" was. And the same was true for the Iraqi soldiers in their desert bunkers, nearly eighty years later. Except that the front was no longer where the main battle occurred. Precisely as called for in AirLand Battle doctrine, the allies were deepening the battle in all dimensions — distance, altitude, time. The front was now in the rear, at the sides, and up above. Actions were planned twelve, twenty-four, seventy-two hours ahead, choreographed in time, as it were.

Long-range air and ground strikes were employed to block or "interdict" the movement of the enemy's follow-on forces, exactly as the Allies had prepared to do in Germany in the event the Soviets ever attacked. The embryonic Third Wave war-form sketched for us almost ten years earlier by Morelli in that hotel room in Crystal City near the Pentagon was no longer a theoretical matter. When the images of war in the Gulf flashed across the world's TV screens, we gasped as we saw more and more of what Morelli, and later Starry, had revealed to us in the early 1980s actually playing itself out in real life in the 1990s.

Destroy the enemy's command facilities. Take out its communications to prevent information from flowing up or down the chain of command. Take the initiative. Strike deep. Prevent the enemy's backup echelons from ever going into action. Integrate air, land, and sea operations. Synchronize combined operations. Avoid frontal attack against the adversary's strong points. Above all, know what the enemy is doing and prevent him from knowing what you are doing. It all sounded very much like AirLand Battle and its updates.

Of course, the Gulf War went beyond AirLand Battle in many respects. Air power played the lead role, rather than its traditional supporting role. So dramatic was this reversal that many concluded air power had at last fulfilled the claims of its early pioneers like the Italian Giulio Douhet (1869–1930), the American Billy Mitchell (1879–1936), and the Briton Hugh Trenchard (1873–1956).

Nevertheless, Iraq was the first full-scale application of updated AirLand Battle doctrine. General Schwarzkopf, the allied commander, reportedly dislikes the term AirLand Battle. If so, it is perhaps understandable. For Schwarzkopf was a brilliant virtuoso performer. However, it takes nothing away from him to say that

Starry and Morelli were the offstage composers who, a decade earlier, wrote the score for the coalition military victory.

Military doctrine is continuing to change in armies around the world. But if we listen closely, whether the words are in Chinese or Italian, French or Russian, the central themes are those of AirLand Battle and AirLand Operations.

When we first met Don Morelli he already understood that changes in the economy and society were also at work in the military. Knowledge, as we've seen, was becoming the key to the production of economic value. What Starry and Morelli did, without necessarily making it explicit, was to place knowledge at the center of warfare as well. Thus Third Wave warfare, as we saw it in the Gulf, shared many of the characteristics of the advanced economy.

When we compare the new features of warfare with those of the new economy, the parallels are unmistakable.

1. FACTORS OF DESTRUCTION

Just as no one would ever entirely discount the importance of, say, raw materials or labor in production, so it would be absurd to ignore material elements in the capacity for destruction. Nor was there ever a time when knowledge was unimportant in war.

Nevertheless, a revolution is occurring that places knowledge, in various forms, at the core of military power. In both production and destruction knowledge reduces the requirement for other inputs.

The Gulf War, writes Alan D. Campen, "was a war where an ounce of silicon in a computer may have had more effect than a ton of uranium." Campen ought to know. He is a retired air force colonel and formerly the Director of Command and Control Policy in the U.S. Defense Department. He now works for the Armed Forces Communications and Electronics Association and is author/editor of *The First Information War,* a highly valuable collection of technical papers on the Gulf War from which some of the data that follows is drawn.

In it, he states, "knowledge came to rival weapons and tactics in importance, giving credence to the notion that an enemy might be brought to its knees principally through destruction and disruption of the means for command and control."

One indicator of the increased knowledge component in warfare is computerization. According to Campen, "Virtually every aspect of warfare is now automated, requiring the ability to transmit large quantities of data in many different forms." And by the end of Desert

Storm, there were more than 3,000 computers in the war zone actually linked to computers in the United States.

On TV, the public saw planes, guns, and tanks, but not the invisible, intangible flow of information, data, and knowledge now required for even the most ordinary military functions. Campen points out, "Most base-level functions are automated on fixed Air Force bases. Supply and maintenance functions are routinely conducted from computers on the flight line."

"At the higher levels of command," writes Maj. T. J. Gibson, an army information specialist, "enemy formations and strengths are tracked and analyzed with computers, courses of action are war-gamed with programs using artificial intelligence, and logistical and personnel information is compiled and tracked on computer spreadsheets."

Over the Gulf flew two of the most powerful information weapons of all — AWACS and J-STARS. Boeing 707 aircraft crammed with computers, communications gear, radar, and sensors, the AWACS (Airborne Warning and Control System) scanned the skies 360 degrees in all directions to detect enemy aircraft or missiles and sent targeting data to interceptors and ground units.

Its counterpart, scanning the ground, was J-STARS — the Joint Surveillance and Target Attack Radar System. It was designed to help detect, disrupt, and destroy the follow-on echelons of an enemy ground force — precisely the task Starry set out to accomplish.

Tipping his braided blue cap to the role played by TRADOC in the development of J-STARS and other key systems used in the Gulf, Maj. Gen. Thomas S. Swalm of the U.S. Air Force says J-STARS provided ground commanders "with a picture of enemy movements as they occurred, as far distant as 155 miles," under all weather conditions.

Two J-STARS planes flew a total of 49 sorties, identified more than 1,000 targets, including convoys, tanks, trucks, armored personnel carriers, and artillery pieces, and controlled 750 fighter planes. Says Swalm, "Aircraft directed by J-STARS had a 90 percent success rate in finding targets on the first pass."

At the same time that the coalition forces were busy collecting, analyzing, and distributing information, they were also busy destroying the enemy's information and communication capability. The Pentagon's final formal report to Congress on the Conduct of the Persian Gulf War — the so-called "COW Report" — points out that the earliest attacks targeted "microwave relay towers, telephone exchanges, switching rooms, fiber optic nodes, and bridges that carried coaxial communications cables." This had the effect of either silencing them, or forcing "the Iraqi leadership to use backup systems vulnerable to eavesdropping that produced valuable intelligence." These attacks were coupled with direct strikes at Saddam's military and political command centers themselves, designed to destroy or isolate the Iraqi leadership and cut it off from its troops in the field.

The task, put differently, was to disrupt the brain and nervous system of the Iraqi military. If any part of the war was "surgical," it was, so to speak, brain surgery.

As understanding of this grows, a recognition is springing up in all parts of the world that a brain-force economy, like that in the United States, Japan, and Europe, implies a brain-based military. Indeed, as we will soon see, even low-tech countries are racing to increase the knowledge-intensive parts of their military.

The flavor of the new thinking is best expressed, perhaps, by Fatima Mernissi, a highly intelligent Moroccan sociologist and feminist, and a passionate Muslim critic of the U.S. role in the Gulf War. "The supremacy of the West," Mernissi has pointed out, "is not so much due to its military hardware as to the fact that its military bases are laboratories and its troops are brains, armies of researchers and engineers."

The day may well come when more soldiers carry computers than carry guns. The U.S. Department of Defense made a start in that direction in 1993 when the U.S. Air Force let a contract for the purchase of up to 300,000 PCs.

Knowledge, in short, is now the central resource of destructivity, just as it is the central resource of productivity.

2. INTANGIBLE VALUES

If, as Starry and Morelli emphasized, seizing the initiative, better intelligence and communications, and better trained soldiers, more strongly motivated, all count for more than sheer numbers, then the military balance may be determined more by intangible, hard-to-quantify factors than by the usual, easy-to-count factors to which Second Wave generals were accustomed.

Just as in the case of obsolete accounting methods in business, military literature is filled with complex, quantitative formulas that attempt to compare forces in terms of their numbers and hardware. The International Institute for Strategic Studies is one of the world's best and most authoritative sources of military data. Its annual, *The Military Balance,* is pored over painstakingly by military planners and media all over the world. It gives detailed information about how many men, tanks, helicopters, vehicles, aircraft, rockets, or submarines are available to each of the world's armies. We, ourselves, have relied heavily on it. But it offers few clues to the increasingly important intangibles. In the future it may tell us how much computing power or communication capability each military enjoys.

In war, just as in business, the ways in which "value" is measured have fallen behind the new realities.

3. DE-MASSIFICATION

When we first met Don Morelli in 1982, he noted that our book *The Third Wave* had introduced the concept of "de-massification."

"But," he told us, "there's one key thing you missed." All this de-massification in the economy and society was going to take place in the military, too. "We are moving," Morelli said, in a memorable phrase, "toward the de-massification of DE-struction in parallel with the de-massification of PRO-duction."

If de-massification in the apparel industry means using a computer-driven laser to cut individual garments, on the battlefield it means using a laser to designate an individual target.

The pharmaceutical industry designs a monoclonal antibody that can identify a disease-causing antigen, enter it through a specific receptor site, and destroy it. The defense industry designs a cruise missile that can identify an Iraqi bunker, enter through its doorway, and destroy it. Smart tools in the economy produce smart weapons for war.

In the civilian economy, advanced technologies sometimes fail. The same is true, of course, for advanced weapons on the battlefield, including the remarkable but controversial Patriot missile. Even the Tomahawk was less than perfect during the war and later, in the 1993 air strike launched by President Clinton against Iraqi intelligence headquarters. Weapons manufacturers routinely overstate the capability of their products. The overall direction of change is clear and indisputable. The goal is finer and finer precision, more and more selectivity.

Built on the same microelectronic base as the civilian economy, smart weapons can detect sound, heat, radar emissions, and other electronic signals, stream this incoming data through powerful analytical software, pick out the identifying "signature" of a specific target, and destroy it. One target, one kill.

To appreciate just how astonishing these new capabilities are it helps to glance briefly backward. In 1881, for example, a British fleet fired 3,000 shells at Egyptian forts near Alexandria. Only ten ever hit their targets.

As recently as the Vietnam War American pilots flew 800 sorties and lost ten planes in an unsuccessful attempt to knock out the Thanh Hoa bridge. Later four F-4s armed with some of the earliest smart bombs did the job in a single pass.

In Vietnam an American M-60 tank crew had to find cover, stop the tank, and aim before it could fire. At 2,000 yards, at night, the chances of hitting a target were, according to tank expert Ralph Hallenbeck, "pretty nil." Today the crew of an M-1 can fire without stopping. Night-vision aids, lasers, and computers that automatically correct for heat, wind, and other conditions assure that they will score a hit nine out of ten times.

Today one F-117, flying a single sortie and dropping one bomb, can accomplish what it took B-17 bombers flying 4,500 sorties and dropping 9,000 bombs to do during World War II, or 95 sorties and 190 bombs during Vietnam.

"What's making all this work," says James F. Digby, a Rand Corporation expert on precision weaponry, are "weapons based on information instead of the volume of firepower. It reduces greatly the tonnage of explosives you have to ship over." His words echo those of business managers who use computers to cut raw material waste and miniaturize products, while slashing inventory and transportation costs.

Mass destruction will no doubt be with us for as long as we can foresee. Weapons will malfunction and deadly errors will continue to be made so long as there is war. But de-massified destruction, custom-tailored to minimize collateral damage, will increasingly dominate the zones of battle, exactly paralleling changes in the civilian economy.

4. WORK

It is by now generally understood that the new "smart" economy requires smart workers, too. As muscle work declines, large numbers of unskilled laborers are increasingly replaced by smaller numbers of highly trained workers and intelligent machines.

This process, too, is perfectly paralleled in the military, where smart weapons require smart soldiers. Poorly educated troops can fight bravely in the hand-to-hand combat that typifies First Wave warfare; they can fight and win Second Wave wars; but they are just as much a drag on Third Wave armies as ignorant workers are on Third Wave industries.

The idea that the Gulf War was a "high-tech" war in which the human element in combat was eliminated is a fantasy. The fact is that the forces sent by the allies to the Gulf were the best educated and technically expert army ever sent into battle. Starry's TRADOC, indeed, trained many of them. It took almost ten years to prepare the American military for the new kind of warfare based on AirLand Battle.

Even advanced armies still have moral Neanderthals in their ranks, as demonstrated by the maltreatment of women during the U.S. Navy's infamous Tailhook convention or by the outbreaks of gay-bashing that still occur. But the changed nature of war places increasing value on education and expertise and less on old-fashioned military machismo and brute force.

The new military needs soldiers who use their brains, can deal with a diversity of people and cultures, who can tolerate ambiguity, take initiative, and ask questions, even to the point of questioning authority. "The 60's slogan 'Question authority' has taken root in the unlikeliest of places," writes Steven D. Stark in the *Los Angeles Times*, describing the changed ethos in the U.S. military. The willingness to ask and think may well be more prevalent in the U.S. armed forces than in many businesses.

Certainly, advanced education today is more common in the military than in the highest levels of business. A recent survey by North Carolina's Center for Creative Leadership showed that while only 19 percent of top American managers had earned a postgraduate degree, a remarkable 88 percent of brigadier generals had advanced education.

Among pilots the levels of training are now far higher than in any earlier period. In World War II, young pilots might be thrown into action after a few hours in the cockpit. Today millions of dollars' worth of training lie behind every F-15 pilot. And it takes years, not days or months, of preparation.

In the words of one U.S. Air Force officer, "The weapons are only as smart as the people using them." Today's pilot is never a solo performer in the cockpit. He is part of a vast, complex interactive system backed up by radar operators in AWACS airplanes to provide early warning of enemy approach, by electronic warfare and counter-warfare experts on the ground and in the air, by planning and intelligence officers, by data analysts and telecommunications personnel. The pilot in his or her cockpit must process vast amounts of data and understand exactly how to fit into this larger system as it changes from instant to instant.

According to two air force colonels, Rosanne Bailey and Thomas Kearney, "The critical factor that leads to success in technology exploitation remains the human element, as typified by the Desert Storm performance of fighter pilots using [the] AIM-7 air-to-air missile. There was more than a five-fold improvement over Vietnam performance . . . the direct result of greatly improved training that emphasized specialized training such as Red Flag and Top Gun [exercises], the use of ultra-realistic simulators that exploit our computer technology, and most importantly, matching the right person with the right job."

The rising educational level is manifest in the lower ranks as well. Over 98 percent of the army's all-volunteer force at the time of the Gulf War were high school graduates, the highest percentage in history. Many were better educated than that. The difference between the conscripted "grunt" in Vietnam and the volunteer soldier in Desert Storm was symbolized for us during the war when we saw a television reporter thrust a microphone into the face of an African-American sergeant standing in front of a tank. The reporter said, "It looks like there's going to be a ground war, soldier. Are you afraid?"

The self-possessed young sergeant looked at him thoughtfully, then replied, "Afraid? No. Perhaps a touch apprehensive."

The careful distinction and the very vocabulary spoke volumes about the quality of the troops. In the words of Marine Col. W. C. Gregson, Military Fellow at the Council on Foreign Relations, the combat arms soldier "is not a mere ammunition mule and bullet hose holder. He understands both mechanized and foot soldier tactics. He is skilled in the operation capabilities of helicopters and fixed-wing aircraft, for he is most often the controlling agent. Directing aircraft means he understands antiaircraft weapons. He is skilled in geometry and navigation, to direct mortars and artillery. . . . Armor and anti-armor, mine and countermine weapons and tactics, use of demolitions, computers, motor vehicles, laser designators, thermal sights, satellite communications gear and organization of supply and logistics are part of his kit." Third Wave combat involves far more than pulling a trigger.

Work force and war-force change in tandem. Mindless warriors are to Third Wave war what unskilled manual laborers are to the Third Wave economy — an endangered species.

We've seen that as economies advance a change occurs in the ratio of "direct labor" to "indirect labor." In the military we see a similar progression.

The military terminology is slightly different. Soldiers speak not of direct or indirect, but of "tooth" or "tail." And the Third Wave tail is now vastly longer than ever before. Notes Gen. Pierre Gallois, "The United States sent 500,000 troops to the Gulf, and there were 200,000 to 300,000 backup troops for logistical purposes. But, in fact, the war was won by only 2,000 soldiers. The tail has grown to immense proportions." That tail even included computer programmers — men and women alike — back home in the United States, some of them working on PCs in their own homes.

Once again what is happening in the economy is reflected in the military.

5. INNOVATION

Another feature of the Gulf War was the high-level initiative shown by troops and civilians alike. "The computer-driven network that fed all-source intelligence to U.S. troops about to plunge across the Saudi Arabian border on February 24, 1991, did not even exist on that day, barely six months earlier, when Iraq invaded Kuwait," says Col. Alan Campen.

It was, he explains, "improvised . . . by a group of innovators who discovered how to bend the rules, end-run the bureaucracy and exploit off-the-shelf hardware and software to get the job done, promptly."

And again: critical systems were put together on the spot by "technicians who, upon discovering that communications and computer equipment would be late in arriving . . . contrived networks by unorthodox and unauthorized use of agglomerations of military and civilian *informationware.*"

Similar stories from the Gulf abound. To a degree unusual in armies, initiative was welcomed — as it increasingly is in smart, competitive firms as well.

6. SCALE

Scale, too, is changing in parallel. Budget cuts in many (though by no means all) countries are forcing commanders to scale down their forces. But other pressures are pushing in the same direction. Military thinkers are discovering that smaller units — like "lean and mean" companies in competitive warfare — can actually deliver "more bang for the buck."

The push is toward weapons systems with more firepower but smaller crews. An experiment being carried out under U.S. admiral Paul Miller, commander in chief of the Atlantic Command, seeks "to assemble troops in smaller, more flexible formations."

Until recently the 10,000–18,000-man division was thought to be the smallest combat unit capable of operating on its own for a sustained period. It would typically include, in the American case, three or four brigades, each with from two to five battalions, along with various support elements and a headquarters staff. But the day is approaching when a capital-intensive Third Wave brigade of 4,000–5,000 troops may be able to do what it took a full-size division to do in the past, and tiny, appropriately armed ground units may do the work of a brigade.

As in the civilian economy, fewer people with intelligent technology can accomplish more than a lot of people with the brute-force tools of the past.

7. ORGANIZATION

Changes in organizational structure in the armed services also parallel developments in the business world. In announcing a recent reorganization, U.S. Air Force secretary Donald Rice explained that a reduced emphasis on nukes and an increasing need for flexible response point to a new structure that enhances the autonomy of the local commander. "The commander of an air base will have unchallenged authority over everything on his facility — from fighters and weather forecasters to radar-jamming planes." Like Third Wave business, the military is loosening its rigid, top-down control.

Perry Smith, a former air force general in charge of long-range planning, became familiar to viewers of CNN when he provided interpretive commentary during the war with Iraq. According to Smith, "Now that the Pentagon has great command, control, and communications facilities, ensuring instantaneous access to our forces around the world, many felt that all wars would be controlled by the Pentagon itself. . . . Yet in the Gulf War just the opposite happened." Field commanders were given a great deal of autonomy. "The central headquarters supported the field commander but did not micromanage him."

This was not only the reverse of how the United States fought in Vietnam. It also contrasted starkly with Soviet practice, which used the new C³I* systems to strengthen top-down authority in a system described as "forward command from the rear."

The downward shift of authority contrasted even more with the way Saddam Hussein ran his army — with commanders in the field afraid to make a move without topside approval. In the Third Wave military, exactly as in the Third Wave corporation, decisional authority is being pushed to the lowest level possible.

8. SYSTEMS INTEGRATION

The growing complexity of the military lends heavier-than-ever significance to the term "integration."

In the air war in the Gulf, airspace "managers," as they are called, had to "de-conflict" the skies — that is, make sure that allied aircraft did not get in one another's way. To accomplish this they had to route thousands of sorties in response to the daily Air Tasking Order. According to Campen, these flights had to move at high speeds through "122 different air refueling tracks, 660 restricted operation zones, 312 missile engagement zones, 78 strike corridors, 92 combat air patrol points and 36 training areas alone, spread over 93,600 miles." All this, moreover, had to be "thoroughly coordinated with the continually shifting civil airways of six independent nations."

The logistics of the war were mind-boggling, too. Even the process of withdrawing U.S. forces after the fighting was a monumental task. Gen. William G. Pagonis was responsible for shipping half a million troops back to the United States. But the task also involved washing, preparing, and transporting over 100,000 trucks, jeeps, and other wheeled vehicles; 10,000 tanks and artillery pieces; and 1,900 helicopters. Over 40,000 containers were moved.

Recently, for the first time, large transport firms have been able, by relying on computers and satellites, to track the packages they carry at every step of the way. Says Pagonis — who not incidentally holds two master's degrees in business administration, "This is the first war in modern times where every screwdriver, every nail is accounted for."

What made this possible for the military were not only computers, data bases, and satellites but their systemic integration.

---

* In the jungle of military acronyms, as in any real jungle, evolution occurs. The ability to command and control troops has been a prerequisite of war since its inception. This led to the abbreviation C², for "Command and Control." As armies came to rely on communications systems to carry orders, C² became C³.

As these systems were integrated with intelligence, the term C³I appeared. And now, as more and more C³I activity depends on computers, the term "Command, Control, Communications, Computers and Intelligence" is giving rise to C⁴I. No end is in sight.

9. INFRASTRUCTURE

Like Third Wave business, a Third Wave military requires a vast, ramified electronic infrastructure. Without it, systemic integration would
have been impossible. Thus the Gulf War saw what has been called
the "largest single communications mobilization in military history."

Starting with minimal capabilities in the region, a complex set of
interconnected networks were built at high speed. These networks,
according to Larry K. Wentz of the Mitre Corporation, relied on 118
mobile ground stations for satellite communications, supplemented
by 12 commercial satellite terminals, using some 81 switches that
made available 329 voice and 30 message circuits.

Extremely complex linkages were established to tie many different
U.S.-based data bases and networks to those in the war zone. In all,
they handled up to 700,000 telephone calls and 152,000 messages per
day, and used 30,000 radio frequencies. The air war alone involved
nearly 30 million telephone calls.

Without this "nervous system" systemic integration of effort
would have been impossible and coalition casualties would have been
sharply higher.

10. ACCELERATION

General Schwarzkopf's famous sweep around the western end of Saddam Hussein's main defenses was a classic application of a turning
maneuver. This "envelopment" was quite predictable to anyone who
bothered to look at a map, although efforts were made to deceive
Saddam Hussein into thinking a frontal attack was imminent.

What was not classic, and what astonished the Iraqi commanders,
was the speed with which the end-run was accomplished. Apparently
no one on their side believed that the allied ground troops could
advance at such historically high speeds. This increase in the velocity
of warfare (like the increasing velocity of economic transactions) was
spurred by computers, telecommunications, and, significantly, satellites.

Unprecedented speed was evident in many other aspects of the
Third Wave war (such as logistics and the construction of the communications facilities). But in contrast, complaints and criticisms surfaced
after the combat that tactical intelligence was too slow in arriving
where it was needed. At the start of Desert Shield, Alan Campen says,

"demands for up-to-date intelligence on the situation in Kuwait and Iraq" were threatening to overwhelm the U.S. Army Intelligence Agency's capacity.

A great deal of information was streaming in from satellites and other sources, but analysis was slow and, lacking adequate communications capacity, photo overlays showing Iraqi ground positions and barrier constructions did not reach the units needing them for twelve to fourteen days. Information produced by the army's Intelligence and Threat Analysis Center still had to be hand-carried to the various corps and divisions in the field via helicopter, truck, and even on foot. These units were spread over a region the size of the eastern United States.

By the time the air campaign began, the delay had been shortened to thirteen hours — a great improvement but still not fast enough.

Many of the systems used to collect and process intelligence were still in their development stage when the fighting began, and some were still in prototype when sent to the Middle East.

But the issue in battle is not necessarily absolute speed, but speed relative to the enemy's pace. And here there was no doubt about the speed superiority of the victors. (Ironically, the intelligence time lags would have been less troublesome if U.S. forces were not themselves moving so quickly.)

Despite these shortfalls, *Forbes,* the business magazine, was right when it wrote, "America won the military war . . . the same way the Japanese are winning the high-technology trade and manufacturing war against us: by using a fast-cycle, time-based competitive strategy."

Of course, a business and an army are decidedly different creatures. No corporate CEO is asked to lay his or her life on the line or to send employees into harm's way. But the way we make wealth is, indeed, the way we make war.

In the Gulf War two military modes, Second Wave and Third Wave, were employed. The Iraqi forces, especially after most of their radar and surveillance were excised, were a conventional "military machine." Machines are the brute technology of the Second Wave era, powerful but stupid. By contrast, the allied force was not a machine, but a system with far greater internal feedback, communication, and self-regulatory adjustment capability. It was, in fact, in part at least, a Third Wave "thinking system."

Only when this principle is fully understood can we glimpse the future of armed violence — and, therefore, of the kind of anti-wars that the future will require.

10

A COLLISION
OF WAR-FORMS

NOW let us briefly set what we have so far seen into the context of past and future.

The idea that each civilization gives rise to its own way of waging war is not new. The Prussian military theorist Clausewitz himself noted that "each age has had its own peculiar forms of war. . . . Each therefore would also keep its own theory of war." Clausewitz went further. Rather than undertake "anxious study of minute details," he declared, those who want to understand war need to make "a shrewd glance at the main features . . . in each particular age."

But at the time Clausewitz wrote, relatively early in the industrial age, there were, as we've seen, only two basic types of civilization. Today, as we've seen, the world is moving from a two-level to a three-level power system, with agricultural economies at the bottom, smokestack economies in the middle, and knowledge-based, or Third Wave, economies likely, at least for a time, to occupy the top of the global power pyramid. In this new global structure, war, too, is trisected.

One predictable result of this will be a radical diversification of the kinds of wars we are likely to confront in the future. It is a military truism that every war is different. But few understand just how varied tomorrow's wars are going to be — and how that increased diversity could complicate future efforts to maintain peace.

To succeed, we will need a better vocabulary to describe the form of warfare that springs from a particular way of making wealth. A century and a half ago, Karl Marx spoke of different "modes of production." Here we can speak of different "modes of destruction," each one characteristic of a given civilization. We can call them, more simply, "war-forms."

Once we start thinking in terms of the interplay of different war-forms, we have a useful new tool for analyzing both the history and future of war.

MACHINE GUNS VERSUS SPEARS

In some wars both sides essentially fight the same way — they both rely on the same war-form. Wars between two or more agrarian kingdoms pockmarked ancient China and medieval Europe. In 1870, to choose another example, France and Germany fought. Both were rapidly industrializing states at roughly similar stages of development.

In another class of wars, war-forms are dramatically mismatched, as in, for example, the colonial wars of the nineteenth century. In India and Africa, Europeans waged industrialized warfare against agrarian and tribal societies. Europe's armies had begun industrializing at least as early as the Napoleonic wars. By the late 1800s they were already beginning to use machine guns (only against nonwhites).

The victors, however, did not conquer vast colonial territories simply because they had machine guns. Backed by societies making the transition from farming to industrial production, their Second Wave armies could communicate faster and better over longer distances. They were better trained, more systematically organized, and had many other advantages as well. They brought a whole new, Second Wave war-form to the killing fields.

In Asia, starting in March 1919, Korean nationalists revolted against Japanese colonial rule. In reminiscing about the 1920s, the man who later became the dictator of North Korea, Kim Il Sung, recalls wondering "whether we . . . could defeat the troops of an imperialist country which produced tanks, artilleries, warships, planes and other modern weapons, as well as heavy equipment, on [the] assembly line."

Adversaries in such conflicts did not simply represent different countries or cultures. They represented different civilizations and different ways of making wealth, one based on the plow, the other on the assembly line. Their respective militaries reflected that clash of civilizations.

A more complicated class of wars pits a single war-form against a dual form. That, as we've seen, is what happened in the Gulf conflict. But it was not the first time an army employed two war-forms at once.

SAMURAI AND SOLDIER

The Europeans had already grabbed huge chunks of Asia when Japan started on its own path to industrialization after the Meiji Revolution in 1868. Determined that it would not be the next victim of European aggrandizement, Japan's modernizers decided to industrialize not merely its economy but its military as well.

Not long after, in 1877, the Satsuma Rebellion broke out. In it, sword-wielding samurai made a last stand against the army of the emperor. The war, according to Meirion and Susie Harries, authors of *Soldiers of the Sun,* saw the last instance of "hand-to-hand combat between individual samurai." But it also saw an early use of the industrial war-form.

While the emperor's force included some First Wave samurai as well, it was largely composed of Second Wave conscripts armed with Gatling guns, mortars, and rifles. Here, as in the Gulf War, therefore, one side relied on a single war-form while the other fought a dual war.

In still another class of wars, which includes within it World War I, we find grand alliances in which both First and Second Wave nations are partners on one or both sides.

Within each of these classes, of course, the wars themselves reflected an immense variety of tactics, forces, technologies, and other factors. But these variations all fell more or less into one war-form or another.

If, however, the past was already marked by considerable diversity, the addition of a Third Wave war-form increases the potential for heterogeneity in the wars we must prevent or wage. The number of mathematically possible permutations shoots up combinatorially.

We already know that older forms of warfare do not entirely disappear when newer ones arise. Just as Second Wave mass production has not disappeared with the coming of customized Third Wave products, so today there are probably twenty countries with regionally significant Second Wave armies. At least some will send infantrymen to die in future conflicts. Trenches, dug-in bunkers, massed troops, frontal assaults — all the methods and weapons of Second Wave war will no doubt continue to be exploited so long as low-tech, low-precision weapons, and "stupid" rather than "smart" tanks and artillery continue to fill the arsenals of poor and angry states.

To make matters more complicated, some First and Second Wave nations now are seeking to acquire Third Wave weaponry, from air defense systems to long-range missiles.

Since in any given year approximately thirty wars of various sizes are raging on the planet, the coming decades could easily see something on the order of fifty to a hundred wars of various sizes as some die down and new ones break out — unless we collectively do a vastly better job of preserving peace and suppressing bloodshed. That task will become more complex as the diversity of wars escalates.

At one end there are small-scale civil wars and violent conflicts in the poor or low-tech world, along with intermittent outbreaks of terror, drug trafficking, environmental sabotage, and similar crimes. But small, essentially First Wave wars at the periphery of the world power system are, as we've argued, not the only type to be feared. The further disintegration of Russia, for example, could throw different mid-tech regions or ethnic groups into Second Wave conflicts using massed forces, tanks, and even tactical nuclear weapons.

High-tech nations on the way to developing brain-force economies could find themselves either sucked into these conflicts or thrown into war as a result of internal political upheavals. Ethnic and religious violence outside their borders can ignite parallel violence inside. Even the possibility of two advanced technological or Third Wave nations fighting one another can no longer be excluded. The air is teeming with trade war scenarios that could translate, if stupidly handled, into actual war between major trading nations.

In short, at least a dozen different mixes and matches of war-forms are possible, each with endless possible variations. And this assumes contests in which there are only two adversaries or simple alliances.

The growing heterogeneity of war will make it vastly more difficult for each country to assess the military strength of its neighbors, friends, or rivals. War planners and war preventers alike face unprecedented complexity and uncertainty. Hyper-diversity also places a premium on coalition warfare (and coalition-based deterrence of war).

In turn when we think about grand alliances involving nations with many different levels of economic and military development, the gradations and varieties skyrocket, as do the potentials for division of labor within coalitions.

Diversity is now raised to so high a level that no country can create an omni-capable military. Even the United States admits the impossibility of financing or waging all kinds of wars. Based on its experience in the Gulf, Washington says that in the future it will seek, wherever possible, to create modular coalitions as crises arise, with each ally sharing in the division of labor by providing specialized military forces and technologies the others might lack. (This approach, incidentally, exactly parallels the efforts of the world's largest corporations to form "strategic alliances" and "consortia" to compete effectively.)

The shift from a bisected to a trisected global power system and to enormously increased military diversity is already forcing armies throughout the world to rethink their basic doctrines. Thus we are in a period of intellectual ferment among military thinkers. Just as the civilization brought by the Third Wave has not yet assumed its mature form, so the Third Wave war-form has not yet reached its full development either. AirLand Battle was only the beginning.

What we have so far seen is, in fact, rudimentary. Originated by the work of Generals Starry and Morelli, revised and later tested on the battlefields of Iraq, the Third Wave war-form is about to be radically broadened and deepened. Widespread cutbacks in military funding, rather than preventing, will accelerate this profound reconceptualization as armies seek to do more with less. A key to that rethinking will be the concept of war-forms and how they interrelate.

A look at changes already under way gives us a startling picture of the nature of both war and anti-war in the early twenty-first century. Unless soldiers and statesmen, diplomats and arms-control negotiators, peace activists and politicians understand what lies ahead, we may find ourselves fighting — or preventing — the wars of the past, rather than those of tomorrow.

PART THREE

EXPLORATION

11

NICHE WARS

EVERYTHING we've seen so far is mere prelude. Even more powerful changes are about to transform wars and anti-wars alike, confronting peacemakers and peacekeepers with strange new questions, some of which will verge on the fantastic.

How should the world deal with endless outbreaks of "small wars" — no two of which resemble one another? Who will rule outer space? Can we prevent or contain bloody wars waged in battlefields crammed with "virtual realities," "artificial intelligence," and autonomous weapons — weapons that, once programmed, will decide on their own when, and toward whom, to fire? Should the world ban — or embrace — a whole new class of weapons designed for bloodless war?

A new war-form does not spring full-blown out of anyone's doctrinal manual, no matter how good. Nor does it come from afteraction studies of a single war. Since it reflects the emergence of a new wealth creation system and, in fact, a whole new civilization, it too emerges and develops as these take form and change the world. Today we can glimpse the trajectory of war itself as the Third Wave warform is extended and deepened.

As we've seen, a Third Wave economy challenges the old industrial system by breaking markets into smaller, more differentiated pieces. Niche markets appear, followed by niche products, niche financing, and niche players in the stock market. Niche advertising fills niche media like cable television.

This de-massification of the advanced economies is paralleled by a de-massification of threats in the world, as a single giant threat of war between superpowers is replaced by a multitude of "niche threats."

Former science adviser to the White House G. A. Keyworth II puts it differently, noting that the shift from highly centralized mainframe computing to "distributed" computing by "hordes of lowly PCs" is paralleled in the "threat environment" facing the global community. Instead of a so-called "Evil Empire," the world now faces "distributed threats."

Changes in technology and economic structure are thus mirrored in warfare as well.

LAUGHING IN THE INFO-SPHERE

Somewhere up in the "info-sphere" where sociologists go when they die, an Italian named Gaetano Mosca is laughing cynically.

Why, he is asking himself, were so many supposedly bright people — politicians, journalists, foreign policy experts, pundits of every variety — shocked or surprised when violence flared up around the world after the end of the Cold War?

"When a war has ended on a large scale," Mosca wrote in 1939 in his book *The Ruling Class*, "will it not be revived on a small scale in quarrels between families, classes and villages?" Mosca, it turns out, was not so far off base — even if the war that ended was cold rather than hot.

Today we see a bewildering diversity of separatist wars, ethnic and religious violence, coups d'état, border disputes, civil upheavals, and terrorist attacks, pushing waves of poverty-stricken, war-ridden immigrants (and hordes of drug traffickers as well) across national boundaries. In the increasingly wired global economy, many of these seemingly small conflicts trigger strong secondary effects in surrounding (and even distant) countries. Thus a "many small wars" scenario is compelling military planners in many armies to look afresh at what they call "special operations" or "special forces" — the niche warriors of tomorrow.

Of all the units in today's armies, special forces or special operations (SO) units probably come closer to waging First Wave war than any other part of the military. Their training emphasizes physical strength, unit cohesion — the creation of strong emotional ties among members of each unit — along with super-proficiency at hand-to-hand combat. The kind of warfare they wage is also the most dependent on the intangibles of combat — intelligence, motivation, confidence, resourcefulness, emotional commitment, morale, and individual initiative.

Special forces — usually volunteers — are, in short, elite units designed, as one officer explained, to function "in areas that are hostile, defended, remote, or culturally sensitive." The term "special operations" covers a vast variety of missions from feeding villagers after a disaster to training the soldiers of a friendly power to fight an insurgency. SO troops may conduct clandestine raids for intelligence gathering, sabotage, hostage rescue, or assassination purposes. They may engage in anti-terrorist and anti-narco actions or wage psychological warfare and supervise cease-fires.

They may move in battalion strength for a commando raid, or in units composed of a handful of men. Recruits are subjected to lengthy training. Says one former special forces officer, only mildly exaggerating, "It takes ten years before you get a truly operational individual. From age eighteen to twenty-eight, he's on a learning curve." Each soldier in a small team is expected to master multiple skills, including fluency in more than one language. Soldiers may get lessons in everything from the operation of foreign weapons to cross-cultural sensitivity.

The May–June 1991 issue of *Infantry* magazine carried an announcement intended to recruit soldiers to "routinely operate throughout the world either individually or in small teams." Those in the know recognized it as a want ad from Delta Force, the U.S. Army's First Special Forces Operational Detachment, designed for hostage rescue tasks. But the Delta Force is only one of the better-known units in the U.S. Army's Special Operations Command. The Navy has its own SO force, as does the Air Force.

On January 17, 1991, even before F-117s made their first strike against Baghdad, three Pave Low Helicopters from the air force's Special Operations Wing led nine army attack helicopters in a streak across the Iraqi border. Flying at thirty feet above the desert, they took out two early-warning radar sites, thus blinding the Iraqis and opening a safe flight path for the hundreds of aircraft that would follow. These were the opening shots of Desert Storm. Other SO troops captured offshore oil platforms held by the Iraqis, conducted deep reconnaissance missions behind enemy lines, performed search and rescue tasks, and carried out other critical tasks.

All told, by 1992 the American Special Operations Command had 42,000 soldiers and reservists in air, sea, and land units. They were deployed in twenty-one countries, including Kuwait and Panama, as well as in Bad Tölz in Germany and Torii Station on the Japanese island of Okinawa.

Similar forces, naturally, exist in many other armies. The former Soviet Union's Spetsnaz troops organized anti-Nazi partisans in World War II. During the Cold War they were tasked to identify and destroy the West's nuclear and chemical weapons and to kill selected allied leaders. Then there is, of course, Britain's famed Special Air Service, or SAS. France's First and Second Parachute Brigades, and its Thirteenth Dragoons Parachute Regiment, are SO forces. Between 1978 and 1991 alone, France sent seventeen military expeditions abroad, mainly composed of these kinds of troops.

Even the smallest of nations maintain such niche warriors, some-times disguised as police, as distinct from soldiers. Denmark has its Jaegerkorps; Belgium its Paracommandos; Taiwan its Amphibious Commandos.

In theory, special forces can be employed in any kind of warfare from a nuclear confrontation down to a tribal border skirmish. But they are especially appropriate in what the military calls "low-intensity conflicts," or LIC — another catch-all term that is applied to hostilities "constituting limited war but short of a conventional or general war."

A LOBBY FOR LIC

Andy Messing, head of the National Defense Council Foundation, is a forty-six-year-old former special forces major who charges into his small, cluttered office outside Washington wearing khaki shorts and an open-necked shirt. He has studied low-intensity conflict firsthand. Having visited twenty-five areas of conflict around the world, from Vietnam and Angola to Kashmir, the Philippines, and El Salvador, he has found himself "up to my ass in combat" in five of them.

Bright and street-smart, Messing is perhaps the most persistent one-man lobby for LIC forces, writing endless op-ed articles in the press, buttonholing members of Congress, lecturing, and hectoring anyone who will listen.

His message is a surprising one — an amalgam of nationalism, populism, and military tough-talk, along with passionate appeals for human rights, action to end poverty and misery in the countries beset by low-intensity war, and theoretical discourses on the futility of waging LIC combat without devoting equal attention to political, social, and economic reform.

Messing sees a world in which many brutal or unstable regimes will be armed with chemical and biological weaponry that may simply have to be surgically excised. The drug war, he says, may have to be expanded. But conflict will also stem from "energy, disease, pollution, and population expansion. . . . I've been to seventeen drug countries," Messing continues. "Peru is drugs. Laos is drugs. But you're also going to see wars in Africa, in places like Zimbabwe or Mozambique because of AIDS."

There are going to be more cases like Somalia or Zaire where governments have broken down entirely and anarchy prevails. Other countries will intervene to protect themselves, to stem the drug trade, to prevent vast refugee flows from crossing a border, or to stop the spread of racial violence across their borders.

This is a world made to order for Third Wave niche warfare rather than the large-scale, total wars of the Second Wave era. As niche warriors proliferate, military doctrine will be adapted to give them added weight. Simultaneously, requirements for new technology will be defined.

Films like *Rambo* that emphasize biceps over brain are already obsolete. The niche warriors of the future will wage information-intensive warfare, making use of the latest Third Wave technologies now on the horizon.

According to the Pentagon's final report on the Gulf War, the initial successful helicopter raid on Saddam's early-warning radar was made "possible because of technological advances in night- and low-light vision devices, precise navigational capability resulting from space-based systems such as the Global Positioning System (GPS) satellites, and highly trained crews."

But these advances only begin to suggest the range of sophisticated technologies already available to special forces. In World War II, Andy Messing says, paratroops might suffer 30 percent casualties just in landing. Their equipment and gear were dispersed over a large area and, often, the soldiers had to fight to join up with one another.

When Iranian radicals took Americans hostage in Teheran in 1979 the United States desperately sought a way to free them. A proposal to land a team of parachutists was rejected for fear they would be scattered over too large an area.

"Today," says Messing, "we have the ability to take a team, jump them from 35,000 feet, twenty-five miles from the objective, at night, the men parasailing down with one eye open and the other looking into an infrared device. They can read a map coming down. They can flash an infrared identification code to one another — one guy flashing two blips a second, another flashing five blips — and they can fly right into a ten-meter area."

FXC Guardian parachutes can provide four feet of forward glide motion for every foot of descent, so a spy or a special forces team can actually be dropped over international waters and glide silently into a country at night, undetected by its radar.

Tom Bumback, a former special forces NCO, now director of operations for a Spec-Ops Expo held near McDill Air Force Base, Florida, tells of a recent demonstration in which a parachutist jumped from 12,000 feet. At 1,000 feet he "cut and ran" — that is, steered — to a touchdown in the Tampa Bay Channel. Plunging below the surface, he swam toward land. Using a "rebreather," he left no bubbles behind. On touching the shore he sprayed the audience with blanks fired from a Calico 5.56 assault rifle, at which point, using a waterproof radio, he called in a helicopter that lowered a line to him and pulled him up 3,000 feet (beyond the range of small arms fire) before reeling him in to safety. "The whole episode from jump to exfiltration took about fifteen minutes," according to Bumback.

When American planes dropped food to besieged villagers in the Balkans, many of the bundles wound up far from their intended landing spots. But today's technology is already obsolete. Thus the AAI Corporation announces recent breakthroughs in airdrop technology. "This is not Buck Rogers," it says. "We've safely dropped 20,000-pound payloads from cargo planes moving at speeds of 150 knots. Each of these drops was completed with astounding, pin-point accuracy.

"This unique system utilizes a cluster of retrorockets that fire as the payload nears the earth, plus a laser altimeter and a sequencer system that tells the rockets exactly when to fire. . . . Soon we'll be able to drop payloads up to 60,000 pounds. Combat vehicles like the Sheridan tank, already assembled and ready to roll."

PH.D. WITH RUCKSACK

Some Special Operations experts are thinking much further into the future. Tomorrow's niche warfare was the subject of a meeting held recently in a small conference room hidden down a winding path at the back of the Old Colony Inn in Alexandria, Virginia.

There approximately fifty listeners — middle-aged businessmen, with a sprinkling of women — leaned forward in their folding chairs as Lt. Col. Michael Simpson of the U.S. Army's Special Operations Command spoke. The audience represented companies, many of them manufacturers of niche products, who sell (or hope to sell) to the army.

Tall and articulate, Colonel Simpson has two master's degrees, one in international relations and the other in strategic studies. But he has also spent fourteen years "hauling a rucksack" in various parts of the world to help carry out "special operations."

His listeners scribbled notes when Simpson began describing his command's future requirements — niche products for the niche conflicts of tomorrow.

Among them were snow and ice vehicles, filmless electronic cameras, lightweight portable power units, chameleonic camouflage (that changes its character as needed), 3-D holographic equipment for training and for combat rehearsals, and automatic voice-translating equipment (American special ops units in the Gulf included two battalions of Arabic-speakers — far too few to meet the need).

Beyond these, Simpson added, "We'd love to have a light, rugged radio unit that combines a global positioning unit, a fax, and on-line coding and decoding capability." Such a device, he said, would "knock thirty pounds off the soldier's backpack."

Another speaker described the need for technologies that could be used for mission planning, threat simulation, training, and rehearsal — all on board an aircraft carrying SO soldiers to their mission. Plan, train, rehearse even while on route to an emergency operation.

SO equipment in general, the suppliers were told, must be simple enough to be used by "indigenous forces," operable under absolute blackout conditions, and have both "LPI and LPD" — a low probability of interception and a low probability of detection.

Col. Craig Childress, a special operations expert in the Pentagon added, "We need vertical-lift aircraft capable of flying horizontally for 1,000 nautical miles," and "we will need to use virtual reality and artificial intelligence" in both rehearsal and actual combat. For example, "Today we have the capability to put a shooter in a room and create a simulated reality we think is real." But in a few years "we should be able to put a whole crew into a simulated reality. The rehearsal should make the actual combat seem like déjà vu. And with AI added to virtual reality we should be able to 'change the reactions of the bad guys' — for example, they might think a door opens to the right when it really opens to the left."

TOWARD MILITARY TELEPATHY

Even more startling possibilities are under consideration. In July 1992 Maj. Gen. Sidney Shachnow of the Special Operations Command presented a "technology time line" projected to the year 2020 that looked toward the development of things like "surreptitiously acquired DNA identification," "whole blood replacement," and even "synthetic telepathy."

Some of these may prove to be no more than fantasies. But other innovations, equally bizarre, no doubt lie ahead. The world needs to start thinking now not just about such technologies but about the future of niche wars generally and the Third Wave war-form of which they are a part.

The deeper implications of Third Wave niche warfare have barely been addressed by governments, peace advocates, or even most military thinkers. What are the geopolitical and social consequences of the rapid development of sophisticated niche war technologies? What happens to the tens of thousands of trained special operations soldiers released into the civilian societies of the world?

Are highly trained teams of Spetnaz troops from the semi-disintegrated former Soviet army marketing their skills to other countries? And what about the thousands of young Arabs and Iranians who poured into Afghanistan from outside to help the mujahedeen combat the Soviets? Many were trained in guerrilla warfare and special operations skills. But their own governments, including Egypt, Tunisia, and Algeria, later made it difficult for them to return home for fear they would put their new skills to work for anti-government revolutionists.

Special forces are military elites. But are military elites, as such, a threat to democracy itself, as some critics insist?

To some, special operations, with their accent on deception, are, in and of themselves, immoral. But so are many of the situations likely to call special forces into action in the fast-approaching future. There is nothing moral about ethnic cleansing . . . cross-border aggression . . . terrorist outrages . . . hostage taking . . . the smuggling of weapons of mass destruction . . . theft of medical supplies and food from humanitarian organizations in the field . . . narco-trafficker bomb blasts, and the like.

SO advocates argue that it is a refined weapon that can be used preventively — to head off a larger conflict, contain small wars, destroy weapons of mass destruction, and for many other positive purposes.

But morality apart, niche warfare will become more important because governments will find it a relatively low-cost option — compared with fielding large conventional forces — to accomplish their goals. It can be used not merely for tactical but for strategic purposes. It may someday be waged not only by governments but international agencies like the UN itself — even perhaps by non-national players on the global stage, from transnational corporations covertly employing mercenaries to fanatic religious movements.

Those who dream of a more peaceful world must put the old nightmares of "nuclear winter" aside and begin thinking imaginatively, right now, about the politics, morality, and military realities of niche warfare in the twenty-first century.

12

SPACE WARS

IN THE FIFTEENTH and sixteenth centuries European powers waxed and waned in their enthusiasm for transatlantic exploration, but once the New World was discovered, there was no turning back. In the same way now, our drive into space may wax and wane, but the competing armies of too many countries are now far too dependent on missiles and satellites to imagine them ignoring the heavens. The vastness of space is a key factor in the war-form of the future.

The Gulf War, writes Col. Alan Campen, former Director of Command and Control Policy at the Pentagon, "is the first instance where combat forces largely were deployed, sustained, commanded, and controlled through satellite communications."

According to Sir Peter Anson and Dennis Cummings of Matra Marconi Space UK Ltd., in Britain, "It was the first real test under war conditions of the $200 billion U.S. space machine, and the first justification in combat of the $1 billion French and British investments in military space."

The earliest U.S. spy satellite was launched in August 1960. By the time of the Gulf War, the U.S. military space "machine" included Keyhole 11 satellites for taking extremely fine-grained photos from space; top secret Magnum satellites for listening into foreign telephone conversations; LACROSSE satellites for collecting radar images of foreign territory; Project White Cloud spacecraft for locating enemy ships; the super-secret Jumpseat satellite for detecting foreign electronic transmissions; plus numerous other communications, weather, and navigational "birds." In all, the coalition made direct use of some sixty allied satellites. Never in history has any army bet so much on events occurring so far beyond the surface of the earth.

THE FOURTH DIMENSION

"Space added a fourth dimension to the war," say Anson and Cummings. "It influenced the general direction of the conflict and saved lives. Space . . . provided detailed images of Iraqi forces and the damage inflicted by allied air attacks. It gave early warning of Scud missile launches. Space provided a navigation system of stunning accuracy that touched upon the performance of every combat soldier, and on missiles, tanks, aircraft and ships." Satellites identified targets, helped ground troops avoid sandstorms, and measured soil moisture, telling Schwarzkopf, the allied commander, precisely what parts of the desert could support tank movements.

Even the small, hush-hush Special Operations units no doubt benefited from space-derived data. Says Ken York, editor of the newsletter *Tactical Technology*, satellites help SO forces "determine the depth of waters for landing parties, potential helicopter landing zones, troop activity, etc." Across the entire military spectrum, therefore, from massive ground movements to the stealthy "insertion" of small paratroop or helicopter-borne teams, space played a crucial role.

Nor will today's round of budget reductions render space any less important. Maj. Gen. Thomas Moorman points out that "Space Command is one of two commands in the United States Air Force that are growing, the other being Special Operations." Says air force Gen. Donald J. Kutyna, chief of the U.S. Space Command, "In a future of decreased, retrenched forces, we will rely on space even more. Space systems will always be first on the scene." This growing emphasis on space changes the entire balance of global military power.

Almost unnoticed by the public and the press, a basic split is widening today between "space powers" and "non-space powers." The latter are collectively asserting that space belongs to everyone and that the benefits of peaceful space activity, irrespective of what country funds them, are the "common heritage" of humanity. Some want to set up a UN Space Agency to control space activities and redistribute the benefits. Battles for the control of space for civilian use will intensify in parallel with its exploitation for military purposes.

Sometimes it will be difficult to separate the two. As global competition heats up, intelligence agencies around the world are focusing more effort on economic and technological intelligence. Military satellite systems that permit countries to listen into, photograph, and otherwise monitor rivals will become weapons in economic as well as military warfare.

But the military significance of space is hardly limited to satellite surveillance. In 1987 there were a total of 850 space and missile launches. Of these, the United States and the then–Soviet Union accounted for approximately 700. All other nations combined shot off only 100–150 launches. By 1989, the worldwide total of launches had doubled to 1,700. Of these, more than 1,000 were conducted by other nations. Put differently, non-superpower launches multiplied ten times within a two-year period.

The fast-expanding list of countries with missiles deployed or in development extends from Iran and Taiwan to North Korea. The missiles vary. Yemen, Libya, and Syria deploy Frog-7s, each with a 70-mile range and the capacity to carry a 1,000-pound warhead. India in 1989 tested the giant Agni missile, which can carry a 2,000-pound warhead 2,500 miles — far enough to hit not only Pakistan, its hostile Muslim neighbor to the north, but Africa, the Middle East, Russia, the former Muslim republics of the Soviet Union, as well as China and many Southeast Asian countries.

With North Korea flooding the Middle East with missiles — and, more important, with factory technology for building more of them — the problem of missile-armed pariah states will grow rather than diminish, and nervousness is mounting. North Korean–built Scud Cs — also called the Rodong-1 — offer customers like Iran longer range, better accuracy, and bigger bangs than the old junkers used by Saddam. While they have a nominal range of 500–600 kilometers, it is believed that certain improvements can actually double those figures. If so, Iran — reportedly in the market for 150 of them –- may now, for the first time, be able to reach out and strike Israel. And North Korea can hit Japan.

All this has encouraged efforts to slow down missile proliferation. In 1987 the G-7 nations — the seven largest economic powers — agreed on a set of common export controls designed to prevent other countries from laying their hands on missiles that could carry a nuclear warhead of more than 227 pounds farther than 175 miles. The agreement is called the Missile Technology Control Regime. But according to Kathleen Bailey, a former official of the U.S. Arms Control and Disarmament Agency, while the MTCR may be of modest help, the fact is that "missile proliferation has unquestionably worsened since the inception of the MTCR" — a fact we will look at in more detail in a later chapter.

FROM IRAN TO ISRAEL

As more countries feel threatened, they begin to think seriously about building or buying their own space surveillance systems to keep an eye on potential adversaries. Even close allies don't want to remain dependent on others for the life-and-death intelligence that satellites can provide.

The defense minister of France has urged Europe to develop its own independent satellite surveillance capability — to make it less dependent on the United States. In turn, the United Arab Emirates decision to buy their own spy satellite from Litton Itek Optical Systems, a Massachusetts firm, provoked sharp objections from some American officials who fear the UAE might share intelligence imagery with other, less friendly, Arab powers. Officials who favor the sale point out that plenty of other nations, like South Korea and Spain, for example, are also contemplating their own systems, and that intelligence satellites are simply going to proliferate whether the United States likes it or not.

MISSILE-PROOFING THE WORLD

On March 23, 1983, President Ronald Reagan proposed the Strategic Defense Initiative, a program aimed at placing a missile-proof protective shield around the United States. This is not the place to review the rancorous decade-long debate that followed. The essential idea, that space-based weapons could shoot down a Soviet ballistic missile before it released its multiple nuclear warheads, was instantly dubbed "Star Wars" by its opponents and ridiculed as unworkable and destabilizing.

With the threat of an all-out U.S.-Soviet nuclear war all but vanished, Reagan's successor President Bush proposed sharply refocusing the program on January 29, 1991. Now it would emphasize protection against accidental or limited nuclear attacks and it would depend mainly on ground-based weapons.

On May 13, 1993, President Clinton's secretary of defense, Les Aspin, announced once and for all "the end of the Star Wars era." In its place a much scaled-down program called Ballistic Missile Defense was announced. The purpose of this was to defend U.S. troops and allies against SCUD-type missiles in regional conflicts like the Gulf War. Further work on space-based weapons was essentially shelved. The underlying assumption of the current shrunken program is that the main threat today comes from short-range missiles in the hands of hostile regimes.

That assumption, however, is itself short-range, if Gen. Charles Horner, head of the U.S. Air Force Space Command, is right. According to Horner, "The technology that went into the SS-25 [a class of huge, mobile and very long-range Soviet missiles] could be available to the high bidders of the world . . . eight to ten years from now." His estimate coincides with that of the CIA, which warned that within a decade at least one Third World country would be able to combine nuclear warheads with missiles capable of striking the United States.

The bottom line is that despite high costs, low budgets, and shrill opposition, pressures for missile defense systems will persist and grow as missiles capable of carrying nuclear, chemical, and biological warheads multiply. (Later we'll look at the chances for stopping the deadly proliferation of such weapons.)

In fact, looking ahead, we can anticipate not just one, but multiple anti-missile systems. It is possible to imagine an Arab version, a Chinese version, even Western European and Japanese systems, if the breach between these countries and the United States is allowed to widen. With North Korea nearby, Japan is racing to upgrade the U.S.-built Patriot system. The British Defence Ministry is studying a limited anti-ballistic missile system to protect the UK against attacks from as far away as 1,875 miles. (Officials point out, as an example, that a Chinese CSS-2 missile based as far away as Libya could hit northern Scotland.) France is pondering a proposal to build its own "anti-tactical ballistic missile system."

Even more striking has been the turnabout in the opinion of the Western European Union, whose members for years remained skeptical about missile defense. In a spring 1993 meeting in Rome, one after another of the European speakers voiced deep concern. Italy's defense minister spoke of "a specific threat for the entire southern flank of Europe" arising from the rapid proliferation of missiles and weapons of mass destruction. Italy, he warned, was "extremely vulnerable" to a military threat nurtured by religious fanaticism, nationalist aspirations . . . and ethnic conflicts." With Libya to its south, with violent Islamic movements menacing governments all across North Africa, with the Balkan war raging right next door to the east, and with Europe itself torn with political and ethnic conflict, his words about Italy's new vulnerability rang loud.

The original idea put forward by President Reagan may be dead, therefore, but with or without Washington, the world is actually gearing up to defend itself against the SCUDs and the larger, more precise missiles of the future.

DROPPING A NUKE ON RICHMOND

Anti-missile defense systems will also refocus attention on anti-satellite weapons (ASATs) designed to take out the eyes and ears of adversaries. In April 1993, even as Congress cut more and more of the Pentagon budget, the chief of staff of the U.S. Air Force made an impassioned speech in which he declared, "We simply must find a way to get on with the construction of capabilities aimed at ensuring that no nation can deny us part of our hard-won space superiority." Urging a complete reconceptualization of American space strategy, he spoke of ensuring that "we can limit our adversaries' ability to use space against us."

To accomplish this, he argued, the United States would need a set of "tools," including anti-satellite capabilities. His words fell on deaf ears and were followed a month later by the forced cancellation of a small army program for an ASAT missile.

The problem facing the United States, however, was not cancelable. "In the Gulf War we faced no attempts to blind or disable our satellites, and our enemy had no access to space for its own purposes. In the not-so-distant future this may change," writes Eliot A. Cohen in the *New Republic*. It is now becoming clear that in the future the first thing any regional power involved in conflict with the United States will do is try to scratch out its eyes in the sky. Ironically, because the United States is the most dependent on its space-based assets and on advanced communications, it is also the most vulnerable to any adversary who can successfully disable or sabotage them.

As early as October 1961 Marshall Rodion Y. Malinovsky, the Soviet defense minister, told Communist Party brass that "the problems of destroying missiles in flight [have] been successfully solved. The following July Khrushchev boasted that Soviet missiles could, in effect, swat a fly in outer space. By early 1968 the Soviets had actually tested an ASAT weapon.

By the mid-1980s they had tested the system against targets in space at least twenty times. Out of a series of fourteen trials, it scored nine kills. By contrast, while the United States could probably deploy an ASAT weapon quickly, it has so far chosen not to, and has actually downgraded work on ASATs. Instead, it has relied on the threat of massive reprisal.

Any direct attack on an American satellite would now be regarded as almost the equivalent of a nuclear attack. As one researcher puts it, "It's not rated quite as bad, perhaps, as dropping a nuke on Washington. But on Richmond, Virginia? Maybe."

SOFT-KILLING THE SATELLITES

To avoid such a face-off, the ex-Soviets and the Americans arrived at a tacit agreement not to shoot at each other's satellites. But shooting a satellite down may be the hard way to blind its owner. It is easier, cheaper, and even more effective to "soft-kill" it — that is, to damage, distort, destroy, or reprogram the information it processes and transmits. There is, in fact, reason to believe that the Soviets once actually did successfully tamper with an American satellite that was later publicly reported to have "died" for mysterious reasons. This was before the two superpowers decided it was too dangerous to play "knock hubcaps" in space.

Some components of U.S. satellite systems are more vulnerable than the public suspects. According to the final Pentagon report on the Gulf conflict, U.S. satellite communications were "vulnerable to jamming, intercept, monitoring, and spoofing [that is, deception], had the enemy been able or chosen to do so."

Worse yet, according to Ronald Elliott, a command and control specialist at U.S. Marine Corps headquarters, as more off-the-shelf components are used in computers and communication networks, it becomes increasingly difficult to detect "unwanted elements" planted in them. Similarly, "mobile satellite networks and wireless computer networks" increase the opportunities for "eavesdroppers and attackers." And as more people design, install, and manage such systems — and as political structures disintegrate or shift alliances — the problems of anti-satellite espionage and brain drains will multiply.

During the Cold War the enemy was known. Tomorrow it may not even be possible to figure out who the adversary is — exactly as is the case with some terrorist attacks today.

BLACK HOLES AND TRAPDOORS

First, potential adversaries are growing more numerous and diverse. Second, methods for sabotaging or manipulating the enemy's satellites and their associated computers and networks are growing more sophisticated. (So-called "black holes," "viruses," "trapdoors," techniques pioneered by hackers to penetrate and cause damage to computer systems, are only the simplest of possible tactics.) Third, it is possible to sabotage an adversary's system while casting suspicion on someone else. Imagine, say, a Chinese attack on American satellite communications disguised as the work of Israeli intelligence — or, for that matter, vice versa. Fourth, it takes very modest physical equipment — much of it available at your local Radio Shack — to manipulate or interfere with satellite signals, ground stations, and their associated networks.

Finally, how do you "massively retaliate" against a terrorist gang or narco-warlord, or even a tiny state, that has no important infrastructure or command center to attack? Or a team of "info-terrorists" arriving in the United States to sabotage critical nodes in the country's highly vulnerable communication system and satellite links. Or, indeed, not arriving at all, but sitting at computer screens somewhere half a world away and penetrating the networks that process and carry satellite-derived data, a problem to which we will return shortly.

After the crack-up of the Soviet Union, the world woke up to the danger that Soviet nuclear scientists, deprived of their jobs and budgets, might sell their dangerous know-how to Libya or Pakistan or other nuke-hungry countries in return for jobs or cash. Are satellite engineers and missile scientists immune to similar blandishments? It doesn't take much to imagine dislocated, disgruntled, and desperate satellite specialists or missileers from, say, the Tyuratam missile test range in Kazakhstan, offering secret information to China. Or to the next Saddam Hussein.

One might even imagine China, for example, with the help of ex–Soviet specialists, learning how to manipulate a whole large subsystem of the ex–Soviet satellite system for its own purposes. For that matter, can we assume that America's "$200 billion space machine" is immune to this kind of manipulation?

Satellite security, moreover, is not just a military concern. Many of the world's most important peace-preserving treaties — treaties restricting the proliferation of nuclear, chemical, or biological weapons; treaties governing troop movements; treaties aimed at building confidence between hostile countries; treaties dealing with certain peace-keeping operations; treaties aimed at preventing ecological warfare in the future — depend on verification of compliance. But a treaty is only worth having if the behavior of its signatories can be monitored. And the chief form of monitoring and verification is surveillance by satellite.

For all these reasons, while no one can know precisely how space war and space-based anti-war will develop in the decades to come, it is clear both will play an even more central role in the twenty-first century.

Before that century ends, unless anti-warriors can get the world to agree to preventive measures, our children may see space rivalries raised to a vastly higher — and more dangerous — level.

THE HEARTLAND IN SPACE

Today no country, including the most advanced, has a comprehensive long-range military strategy for space. This point is made by John Collins, author of an extremely important but largely unknown study that analyzes the entire earth-moon system in military terms. Commissioned by the United States Congress and entitled *Military Space Forces: The Next 50 Years*, the book deserves close reading.

Collins, a senior analyst at the U.S. Library of Congress, cites the geopolitician Halford J. Mackinder (1861–1947), who at the turn of the century developed the theory that East-Central Europe and Russia comprised the "Heartland" of global power. Africa and the rest of Eurasia were merely the "World Island."

Mackinder formulated a much-quoted rule that ran as follows:

- Who rules East Europe rules the Heartland.
- Who rules the Heartland commands the World Island.
- Who rules the World Island commands the World.

Nearly a century has passed, and Mackinder's theory is no longer taken seriously because air power and space power have made turn-of-the-century geopolitical assumptions obsolete. But Collins draws a dramatic analogy from Mackinder. "Circumterrestrial space," he explains, ". . . encapsulates Earth to an altitude of 50,000 miles or so." And that, he suggests, will be the key to military domination by the mid-twenty-first century.

- Who rules circumterrestrial space commands Planet Earth.
- Who rules the moon commands circumterrestrial space.
- Who rules L4 and L5 commands the Earth-Moon System.

L4 and L5 are lunar libration points — locations in space where the gravitational pull of the moon and the earth are exactly equal. In theory, military bases planted there could stay in position for very long times without needing much fuel. They may be the equivalent of "high ground" for the space warriors of tomorrow.

Right now all such talk has a tinny, sci-fi ring to it, but so did early forecasts about tank warfare or air power. Anyone who brushes ideas like these entirely aside, or who thinks the drive to exploit space for military purposes is over, or that budget cuts will lay it to rest, is being myopic.

Not just Third Wave war, but Third Wave anti-war as well will increasingly depend on actions beyond the earth. Preventive peace-making requires us to peer beyond the present. What is at issue is not simply dollars; it is human destiny.

13

ROBOT WARS

MEDIEVAL Jewish legend told of an automaton called the
"Golem" that mysteriously came to life to protect its owner. Today a
new breed of "Golem" is on the horizon — robot warriors — and no
serious look at Third Wave war and anti-war can afford to ignore
them.

Talk about robots on the battlefield is old and cheap. Ever since
World War I the attempt to build practical military robots has run
into one snag and embarrassment after another. The uninformed public still associates fighting robots with science fiction movies like *Ro-
bocop* or *Terminator 2*, and traditionalist officers remain skeptical.

Nonetheless, military thinkers around the world are taking a fresh
look at this technology. New conditions, they say, will lead to a
stronger-than-ever push for robotization. Lewis Franklin, formerly
vice president of the Space and Defense Sector of TRW, a leading
defense contractor, believes we can expect a mini-flood of robotic
systems entering into military life in about ten or fifteen years.

To choose a single example, the Gulf conflict gave remotely piloted
vehicles, or RPVs, a major boost. According to *Defense News,* the
war "galvanized support" for these aircraft, so that "international demand for pilotless combat aircraft is expected to explode."

Manufacturers of all kinds of military robots expect a $4 billion
market before the decade is out, in spite of defense cutbacks. They
look for U.S. spending for robots to rise tenfold. Whether that optimistic forecast is met or not, says Lt. Joseph Beel, a faculty member at

the U.S. Naval Academy, other countries may well use them against U.S. forces in future conflicts.

Various long-term factors lend credibility to these forecasts. The first is purely technological. As robots proliferate in both factories and offices, civilian research on robotics is advancing swiftly. From chips that control "self-healing" phone networks to "intelligent buildings" and "smart highways," a technical base is being laid for speedier robotization of the economy in the future. This will, in turn, spin off a host of applications with military potential.

A BATTLEFIELD BARGAIN

In civilian economies in which labor is cheap, the advance of robotization is slow or nonexistent. As labor costs rise, however, automation in general and robotization in particular become competitively advantageous. Much the same applies to armies. Poorly paid draftee armies reduce the incentive for technological substitution. By contrast, if armies consist of more highly paid professionals, robots become a battlefield bargain.

The spread of chemical, biological, and nuclear arms in the world is also likely to promote robotization by creating battlefields just too toxic for human soldiers. Robotic warriors can be custom-designed to perform in just such environments.

But the most important factor favoring robotization may well be the change in the public's attitude toward "acceptable" casualty levels. According to Maj. Gen. Jerry Harrison, former chief of the research and development labs of the U.S. Army, the extremely low allied losses in the Gulf War "set a standard that surprised many people. To replicate that in the future war translates into robotics."

Among the most hazardous duties in battle are helicopter reconnaissance and scouting missions. One way to cut down chopper casualties, for example, would be to launch fleets of low-altitude robots the size and shape of model planes, each with a specialized sensor of a different type, each feeding data back to the field commander. According to *Strategic Technologies for the Army of the 21st Century*, a report prepared by the U.S. Army after the experience of the Gulf War, such drones provide "a less vulnerable, less costly alternative that does not risk crew lives."

Henry C. Yuen has another idea. (Yuen is perhaps best known as the inventor of the VCR *Plus+* device that makes it possible to program your VCR without first taking a degree in electronic engineering.

Creating that was a sideline, however, while Yuen, an expert in anti-submarine warfare, worked at TRW.) In an internal paper written shortly after the war in the Gulf ended, Yuen argued that "one of the foremost objectives in the development of new weaponry should be the reduction or total elimination of human risk. Put simply, weapons or equipment in harm's way should, to the extent possible, be unmanned," that is, robotic. Yuen outlined plans for driverless tanks that would operate in teams under the control of a remote battle station.

PROTECTING THE A-TEAM

The same ideas are echoed by General Harrison: "You protect your A-team, you protect your varsity squad — your soldiers, your pilots — until you absolutely have to put them into the conflict. And you do that by using a robot. . . . I could have a centrally controlled tank and have six followers that don't have any people in them. One guy controls six of them robotically."

Franklin, Yuen, and Harrison are only a few of the many voices now advocating rapid robotization. Robots could do more than replace reconnaissance helicopter pilots or tank drivers. In addition to gathering intelligence and spotting targets, they can be used to deceive or destroy enemy radar, to collect data about damage inflicted on an opponent, to repair equipment and patrol perimeters. A long list of other uses are also possible. They run the gamut from recovering and defusing live warheads to providing logistical support, cleaning up toxic environments, planting sensors under soil or sea, clearing mines, fixing bomb-damaged runways, and beyond. Harvey Meieran of PHD Technologies Inc., in Pittsburgh, in a paper delivered to a recent conference of the 2,500-member Association of Unmanned Vehicle Systems, cited at least fifty-seven different combat functions that robots can perform.

Military roboteers are naturally pleased about the new respect being shown toward their work. They are also excited by the promise held out by recent advances in artificial intelligence, virtual reality, computer power, display systems, and related technologies. But they are torn by controversy about what happens next. The question that agitates them is not how to make robot weapons clever, but how clever to let them become.

A quiet debate is taking place among these engineers that raises some of the largest issues facing the human race. The issue concerns not simply war or peace, but the possible subordination of our species to super-intelligent, increasingly self-aware killer-robots.

ROBOTS OVER THE DESERT

Long the province of science fiction pulp magazines and movies like *The Forbin Project*, robots that actually think for themselves (or mimic thought) are now, for the first time, being taken seriously by the men and women who are designing the war technologies of the not-too-distant future. An ideological conflict has developed between backers of "human-in-the-loop" robots and advocates of "autonomous" weapons that are intelligent enough to act on their own.

While robotic weapons played only a small part in the Gulf War, the most evident were under human control. The skies over Kuwait and Iraq were dotted with Pioneer RPVs — small, unarmed, pilotless planes — under the control of "tele-operators" sitting at computer consoles miles away. Robots did the work, but humans made the decisions.

Designed by Israel and built by a U.S. firm, the Pioneer "drones" went almost unnoticed by the media — not to mention by the Iraqis. Some were launched from the deck of the battleship USS *Wisconsin*, others by U.S. Army and Marine Corps ground units. According to Edward E. Davis, the navy's deputy program manager for "unmanned aerial vehicles," Pioneers flew 330 sorties and spent over 1,000 hours in the air once Desert Storm began. One remained airborne twenty-four hours a day throughout the entire period of combat.

These RPVs performed reconnaissance missions, checked on bomb damage, searched for mines in the Gulf, watched for Iraqi patrol boats, and carried out other tasks. Three were hit by small arms fire. One was shot down.

Pioneers in the air tracked Iraqi mobile missile launchers as they returned to their bases, spotted Silkworm missile sites and determined whether they were active or inactive, and observed Iraqi ground forces massing for the brief, ill-fated attack they made on al-Kafji in Saudi Arabia. Information collected by cameras or sensors on the pilotless planes were fed to ground stations and then to the Cobras and Av-8Bs that flew out to strike the Iraqi formations. Elsewhere

Pioneers reconnoitered routes and determined the flight plans to be followed by the army's Apache helicopters.

Nor were the Pioneers the only "human-in-the-loop" robots used. The U.S. Eighty-second Airborne deployed an experimental Pointer drone that can be carried in two backpacks and assembled in five minutes. It was used to patrol base perimeters. Other unmanned aerial vehicles, including the Canadian CL-89 and the French-made MART, were used as well for identifying targets, as decoys, or for other functions. Nor were robots limited to air operations. German minesweepers reportedly deployed unmanned patrol boats called TROIKAS.

RETRACT MAPLE

Experiences like these have spurred work on far more ambitious projects. The U.S. Navy is spending over half a billion dollars on a secret program called Retract Maple that will permit a commander on Ship One to receive radar and other instantaneous information from Ship Two and to fire missiles automatically from ships Three, Four, or, for that matter, Ten or Twenty. Retract Maple can also send out decoys and jam the guidance system of incoming enemy missiles. It gives the task-force commander remote control over an entire task force consisting of a large number of ships, from cruisers and destroyers on down.

By extension, one can envision even more complex integrations of helicopters, ships, tanks, and ground-support planes into a single "robotic organism" under the control of tele-operators. The imagination conjures up an all-robotic battlefield.

Today literally hundreds of different robotic R & D projects are under way, from Italy and Israel to South Africa, the Russian republic Germany, and Japan. But even those presumably designed for civilian purposes may create "dual" technologies.

Japan Aviation Electronics Industry Ltd. has built a remotely controlled helicopter that, in the words of JAEI's Toshio Shimazaki, could be used "to take pictures and collect data on temperatures, emissions or other factors near tanker fires or submarine volcanoes." Yamaha, known for its pianos and motorcycles, has developed the R-50 remotely controlled helicopter for crop dusting. Kyoto University and two government agencies are building a small robot plane with potential meteorological, environmental, and radio transmission functions. It is designed to stay aloft indefinitely with power supplied by microwave from below. Komatsu Ltd., meanwhile, has created a multi-legged robotic device for use in underwater construction.

Japan's constitution bans the export of arms. But one wonders what would prevent such an underwater robot from being used to plant mines or sensors in otherwise inaccessible places? Indeed, all these robots — exactly as with trucks or jeeps — can be used for military as well as purely civilian purposes.

Many robots are tailor-made for protecting factories — not to mention missile bases or nuclear facilities — against terrorists. Perhaps the best overview of military robotics is a short book called *War Without Men*, by two researchers, Steven M. Shaker and Alan R. Wise. According to Shaker and Wise, from whom many of these examples are drawn, Robot Defense Systems, a Colorado company, has created a two-ton wheeled vehicle called the Prowler for sentry duty.

THE PROWLER

The Prowler can be operated from a distance of nineteen miles away. Stuffed with computers and swiveling video cameras, the device can circle an installation or keep watch on its entry point. It uses laser range finders and other instruments to position itself, including sensors that indicate changes in the terrain it traverses. The operator at a distance "sees" what the cameras find as they scan.

The vehicle can be fitted with night-vision equipment, infrared scanners, radar, and electromagnetic motion sensors and seismic detectors. It can also be equipped with a wide variety of weapons. The giant contracting firm Bechtel National, we are told, has proposed its use "for security work at an installation in a Middle Eastern country."

Meanwhile, Israel, surrounded by hostile neighbors, with its army vastly outnumbered by its enemies, has become a world leader in the design and application of robotic technology to both peace and war. Not far from the Sea of Galilee, the Iscar plant manufactures cutting tools for export. Built by a high-tech visionary named Stef Wertheimer and his son Eitan, the plant is a world-class model of factory robotization. The military use of robots is also highly advanced in Israel, which employed RPVs with spectacular success against the Syrians in Lebanon in 1982 and has used them in antiterrorist actions as well. In one case a remotely piloted plane followed a car carrying fleeing terrorists back to their base, so that it could subsequently be demolished by air attack.

ROBO-TERROR

However, as Shaker and Wise point out, "terrorists are becoming more sophisticated in countering robotic technology." They cite a case in which a robot, under the control of a remote operator, was being used to defuse a bomb. Revolutionaries were "able to override the . . . operator's radio control and have the robot turn on him. The operator barely escaped being blown up by his own robot."

And they continue, "Robotic vehicles with no moral conscience, and without any fear of suicide missions, might . . . make ideal terrorists. The use of mechanized killers would certainly cause panic and concern among victims and generate the publicity sought by terrorists."

So far, we have been talking about "human-in-the-loop" robotry. But these are only a first, half step in the march toward the more advanced — and far more controversial — autonomous robots. Compared with these, remote-controlled or tele-operated robots are only semi-smart. There are smarter devices, like the Tomahawk cruise missile, which, once launched, no longer receives instructions, but is preprogrammed to behave independently.

The final step are weapons that, once "born" or set in motion, make more and more of their own decisions. These are the so-called "autonomous" weapons and, ultimately, says Marvin S. Stone, general manager of TRW's electronics and technology division, "All of the weapons will be more autonomous."

The problem with remotely controlled robotic weapons is that they depend on vulnerable communications that link humans to less bright, but nicely responsive mechanical extensions of themselves. If communication breaks down, or is disrupted, or sabotaged, or, worse yet, manipulated by the enemy, the robot becomes useless or potentially self-destructive. If the ability to sense data, interpret it, and make decisions is embedded in the weapon itself, the communication links are internalized and more secure.

Another feature of autonomous robots is speed. They can make decisions at faster-than-human rates, a key capability as warfare accelerates. Shaker and Wise point out that the various parts of a missile defense system "must exchange data at such high rates of speed to counter a strategic attack that humans will be unable to participate as 'on-the-spot' decision makers."

If robots can be trusted with making such decisions autonomously, they had better be super-smart. Hence, the search for robots that can actually learn from their own experience. The U.S. Naval Research Laboratory has developed software that, according to *Defense News*, "allows robot vehicles to make rudimentary judgments and learn by coping with unexpected circumstances." Tested in a flight simulator, the software learned to land an F/A-18 safely on the heaving deck of an aircraft carrier 100 percent of the time. The same software was able to increase the plane's ability "to evade anti-aircraft missiles from 40 per cent of the time to 99 per cent."

Advocates of autonomous weaponry thus claim that they offer superior security, speed, and, in some cases, the ability to learn from their own experience. What's more, like tele-operated robots they can be linked together to form giant systems.

As originally conceived, the Strategic Defense Initiative, with its worldwide network of satellites, sensors, and ground stations, could be seen as a single autonomous "mega-robot," at least some parts of which would operate autonomously. But even these plans barely scratch the surface of possibility.

Quite apart from SDI, DARPA, the Defense Advanced Research Project Agency at the Pentagon, began supporting research into self-deciding vehicles a decade ago. Its SHARC program has looked at what might be done by an entire group of inter-communicating robot vehicles. One can even imagine a kind of collective "consciousness" or quasi-telepathy emerging among them.

THE ANTI-ROBOTEERS

This, perhaps, helps explain at least some of the resistance faced by the roboteers. Here again there are parallels with the civilian economy. Exactly as in the business world, military robotization looms as a threat to vested interests. Once again, Shaker and Wise: "In the factory it is the blue-collar workers whose jobs face extinction by automation. . . . In the military . . . oftentimes the upper management are hands-on operators of weapons systems; many of their roles are at stake because of the potential introduction of robotic vehicles. Their resistance is likely to be fiercer than what has occurred in the factories."

They point out that in the United States "the air force flag rank is largely made up of pilots. In the navy, both aviators and ship commanders are in control of the organization. In today's army, command accrues primarily to those associated with combat soldiers. It is the same in other nations' military establishments. Planners, intelligence officers, communications officers, acquisition managers, and other noncombat specialists rarely reach the pinnacle of power." The shift to Third Wave warfare, and most particularly the shift toward robotization, could change all that, cutting the perks and power of the officers now managing manned systems.

Yet the case against robotics — and especially autonomy — cannot simply be dismissed as self-serving. Anti-roboteers argue that robotic weapons can't adapt to the myriad sudden changes on the battlefield. Can human overrides be built in at every step? What is the morality of robot-killers who may not be able to distinguish between an enemy who is a threat and one who is trying desperately to surrender? Could malfunctioning robot weapons go haywire and trigger an endlessly escalating conflagration? Are human programmers smart enough to anticipate every possible change in battlefield circumstance?

This, then, is precisely where the Dr. Strangelove scenario begins. By taking the human out of the loop, don't we risk runaway war? Pro-roboteers can point to the fact — little known by the public — that some of our deadliest nuclear weapons systems are, and have long been, dependent on partially autonomous components. The speed and danger associated with a nuclear attack by the Soviets were both so great that only by relying on a certain degree of autonomy could deterrence be assured. And despite this fact, no accidental or runaway release of nuclear weapons has occurred since the dawn of the nuclear age over half a century ago. Human decision makers, it hardly needs saying, can also go haywire.

Not everyone, however, is reassured. The difference, it is said, is that if humans go gaga, there may be time to stop them or to limit the consequences of their decisions. That may not be the case if we endow robotic weapons systems with suprahuman intelligence, give them the power to make instantaneous choices, to learn, and to communicate with one another.

Even the best robotic designers can and do make mistakes. Even the finest software team cannot "think of everything." The danger is a failure to be fail-safe, an inability to cope with error, surprise, and chance — precisely the phenomena that proliferate in what Clausewitz called "the fog of war."

Such dismal considerations have led distinguished computer scientists to oppose military robotization altogether. But the reality is less black and white. There are an almost infinite number of possible mixtures — systems that combine tele-operation with varying degrees of autonomy. And it is these that seem likeliest to proliferate in the early twenty-first century. Robots, like satellites and missiles and high-tech niche warfare, will, whether we are ready for them or not, take their place in the emerging war-form of Third Wave civilization.

Carried to its ultimate, the debate over autonomous weapons pushes us beyond the beyond. If work at far edges of military robotics ever were to converge with the research under way in the field of computational biology and evolution, all current bets would truly be off. In the T-13 Complex Systems Group at the Los Alamos National Laboratory, A-Life researchers study human-made systems that mimic living systems which evolve and develop the capacity for independent behavior. The scientists in this field worry endlessly about its moral and its military implications. Doyne Farmer, a Los Alamos physicist who has since left to form his own firm, speculated in an essay co-written with Alletta D'A. Belin, that "Once self-reproducing war machines are in place, even if we should change our mind . . . dismantling them may become impossible — they may literally be out of our control."

In the next chapter we will meet some "self-reproducing war machines." But long before these become available a question needs to be asked: how, and to what degree, can all the concentrated human imagination and intelligence invested in robotry be applied to peace as well as war? Can robotics contribute as much to Third Wave anti-war as to Third Wave war?

14

DA VINCI DREAMS

LONG before Leonardo da Vinci began toying with the idea of flying machines and fantastic forerunners of the tank, the rocket, and the flamethrower, creative minds conjured up weapons of the future.

Today, despite cutbacks in military spending in many (though by no means all) countries, military imagination is still hard at work. If we ask thoughtful military men what their forces will need in the years ahead, they pull out of their desk drawers a dazzling list of dream weapons. Few of these will ever actually come into being. But some of them *will* materialize and play their part in Third Wave warfare.

What most nations now want are smarter weapons, beginning with sensors. American military planners hunger for next-generation sensors able to detect fixed and moving objects from 500 to 1,000 miles away. Such sensors would be mounted on aircraft, drones, or space vehicles, but more important they would be under the decentralized control of theater commanders, who would be able to move them around as needed and customize the information streaming in from them. This smart sensor of the near-term future would bring together or "fuse" different kinds of fine-grained data, synthesize it, and check it against many kinds of data bases. The result would be better early warning, more refined targeting, and improved damage assessments. Sensors are top priority.

On the ground, the army wishes to replace stupid, inert mines with smart mines that don't wait for an enemy tank to roll over them. Instead the "dream mine" would acoustically scan the area around it, compare engine sounds and earth rumbles against a list of vehicle types, identify the target, use an infrared sensor to locate it, and then fire a shaped charge at it.

The U.S. Army is also looking into "smart armor" for its own tanks. As an incoming projectile approaches, a mesh of sensors mounted outside the skin of the tank would measure and identify the type of round and instantly communicate that information to an on-board computer. Tiny explosive "tiles" on the outside of the tank would be fired off by the computer to deflect or destroy the inbound shell. Such advanced armor would fend off either kinetic or chemical warheads.

Other planners picture an all-electric battlefield, spelling the end of the Age of Gunpowder for artillery. In this scenario electricity propels the shell and electronics guides it to its target. All vehicles are electrical, recharged, perhaps, by aircraft that fly over them and zap energy to them.

A HOLLYWOOD SUIT

The individual soldier is also reconceptualized. According to Maj. Gen. Jerry Harrison, former head of research and development laboratories for the U.S. Army, the soldier should no longer be viewed "as something you hang a rifle on, or that you hang a radio on, but as a system."

Already under research is the concept of SIPS — the Soldier Integrated Protective Suit. This is a "suit" that would offer protection against nuclear, chemical, or biological weapons, provide the soldier with night-vision goggles and a heads-up display. It would also include an aiming system that tracks eye movements so that it can automatically point the gun at whatever the soldier is looking at.

These and additional capabilities would all be integrated into a suit that is right out of a Hollywood special effects department — an intelligent exo-skeletal suit that learns to perform the soldier's repetitive tasks so he or she can march ten miles and doze off while doing it . . . a suit that amplifies the strength of the wearer several-fold. As General Harrison puts it, "I want to put this guy in some sort of exo-skeletal suit that will allow him to leap tall buildings with a single bound." The allusion to Superman is clear.

The soldier inside this smart suit, however, is not an overmuscled, small-brained cartoon character but an intelligent man or woman capable of processing huge amounts of information, analyzing it, and taking resourceful action based on it.

This vision of every soldier a Superman or Schwarzenegger, or, more accurately a Terminator, is taken seriously enough for a group of researchers to have formed around the concept at the U.S. Army's Human Engineering Laboratory in Aberdeen, Maryland.

According to Maj. Gen. William Forster, director of combat requirements in the Pentagon, the ultimate object of the work on SIPS is "to increase the effectiveness of the individual so that you need fewer soldiers. The fewer 'soft-skin' soldiers we have out there, the fewer the casualties."

Science fiction–like or not, Forster notes, "The Exo-Skeleton or Exo-Man is being widely discussed, and even though it is far-out, all these things are within the known laws of physics. You don't have to change the laws to do them. The real trick is doing them economically and reliably."

AN INFESTATION OF "ANTS"

Also within the framework of known laws are even more remarkable possibilities. Micro-machines, for example. Today the first micro-machines are just being patented — for example, an electric motor less than a millimeter long that could, according to Prof. Johannes G. Smits, drive a robot the size of an ant.

"Imagine what you could do with an ant if you could control it," says Smits, an electrical engineer at Boston University, who holds the patent on the new motor. "You could make it walk into CIA headquarters." The energy to drive the micro-robot could come from a micro-microphone that converts sound into energy.

It doesn't require much imagination to appreciate what an infestation of robotic ants could do to an enemy's radar installation, or to aircraft engines or to a computer center.

Such micro-machines, however, are huge, hulking giants compared with the nano-machines to come. If micro-machines are small enough to manipulate individual cells, nano-machines can manipulate the molecules of which cells are built. Nano-robots would be small enough to operate like submarines in the bloodstream of humans, and presumably could, among other things, be used to perform surgery at the molecular level.

Work on nano-technology is under way in the United States and Japan, where researchers Yotaro Hatamura and Hiroshi Miroshita have prepared a study on Direct Coupling Between Nanometer World and Human World. According to a survey of twenty-five scientists working on nano-tech, within the next ten to twenty-five years we will not merely be able to create devices at the molecular scale, but will be able to make them self-replicating — meaning we can breed them.

Here we approach the "self-reproducing war machines" alluded to earlier. For example, the smart sensors we have been talking about so far are near-term extensions of current technology. But a generation from now, says a physicist at the RAND Corporation, "we start looking at sensors that . . . can burrow into communications systems, or sensors that can lie there for twenty years, just ticking away, ready to be remotely activated. They could be the size of a brilliant pinpoint under the ground."

Imagine, then, super-smart sensors and mines, a few nanometers in size, that can, as suggested in the preceding paragraphs, reproduce themselves. Now picture a scenario in which a global police force seeds them over a pariah state and programs them to replicate to a given density in militarily sensitive regions. Virtually undetectable and harmless, the mines could be armed selectively from the outside by tiny pulses of energy. At which point the local Saddam Hussein is told to close down his chemical weapons plant or see all his military bases erupt. Unless, of course, the enemy reprograms them. Or they refuse to stop breeding. Of course, all this is, at this point, just fantasy. But so were Leonardo's flying machines when he drew them.

SUPER-PLAGUES

We need not wait for self-breeding nano-technology, however, to face novel terrors. Long before then the diffusion of swiftly advancing scientific knowledge threatens to turn conventional chemical and bio- logical weaponry into the so-called "poor man's nuclear bomb."

While it still remains cumbersome to handle and deliver most chemi- cal or biological weapons without endangering one's own forces, that is hardly likely to inhibit the Pol Pots or Saddam Husseins of tomorrow. The world has justifiably begun to worry about chemical and biological weapons programs in countries like Libya, India, Pakistan, China, and North Korea, not to mention Iraq, many of which may face political and economic instability in the decades to come.

In January 1993, with much self-congratulation, after a quarter century of negotiation, 120 nations met in Paris to sign the Chemical Weapons Convention. Theoretically it bans the production and stor- age of chemical arms. A matching body, the Organization for the

Prohibition of Chemical Weapons (OPCW), was established to police the agreement. Its inspectors will have greater powers than those enjoyed by the International Atomic Energy Agency (IAEA) until now. But twenty-one members of the Arab League refused to join in the agreement until Israel did. Iraq sent no one to the meeting. And the Convention actually does not come into effect until half a year after fully sixty-five nations ratify it.

Even Russia, which has sworn up and down to eliminate chemical arms, recently arrested two scientists, Vil Mirzayanov and Lev Fyodorov, for revealing in the press that a new chemo-weapon was being developed at a lab in Moscow after Russian president Yeltsin had spoken out in favor of agreements with the United States to get rid of such toxins.

As for biological warfare agents — in many ways the worst of the weapons of mass destruction — it is now known that work on offensive bio-war weapons continued in the Soviet Union long after it signed a 1972 treaty outlawing such arms; long after these activities were denied by Gorbachev; long after the Soviet state collapsed and was replaced by Russia; and even after Yeltsin publicly ordered germ warfare research ended. That work included — and may still include — a search for a genetically engineered "super-plague" that could wipe out half the population of a small city in short order.

Who, in a country torn apart politically and on the edge of anarchy, controls the pathogens that still, no doubt, remain in the laboratories of the former Soviet Union? And how safe are they?

In 1976, the Soviets, undoubtedly aware of the horrors breeding in their own laboratories, called for international bans on exotic arms. They warned, at that time, of the hideous possibility of race-specific weapons — genetically engineered to single out and decimate only the members of selected ethnic groups — the ultimate genocidal weapon for race war. In 1992 Bo Rybeck, director of the Swedish National Defense Research Institute, pointed out that as we become able to identify the DNA variations of different racial and ethnic groups, "we will be able to determine the differences between blacks and whites and Orientals and Jews and Swedes and Finns and develop an agent that will kill only [a particular] group." One can imagine the uses to which such technology might be put by "ethnic cleansers" of tomorrow.

The warning about race-specific weaponry takes on new urgency in light of recent scientific advances connected with the Human Genome Initiative, which aims at unlocking the secrets of DNA. Taken a

step further, it conjures up the use of bioengineering or genetic engineering to alter soldiers or to breed "para-humans" to do the fighting.
Fantastic, no doubt. But no longer beyond the extremes of possibility.

And then there is ecological weaponry. When Saddam Hussein
torched the Kuwaiti oil fields, he was only doing what the Romans
did when, according to some, they salted the fields of Carthage, and
what the Russians did to their own fields during World War II when
they pursued their "scorched earth" policy to deny food to the Nazi
invaders. And, indeed, what the United States did with the use of
defoliants in Vietnam.

These acts are primitive compared with some of the imaginable
(and imagined) possibilities of sophisticated ecological weaponry. For
example, triggering earthquakes or volcano eruptions at a distance by
generating certain electromagnetic waves; deflecting wind currents;
sending in a vector of genetically altered insects to devastate a selected
crop; using lasers to cut a custom-tailored hole in the ozone layer
over an adversary's land; and even modifying weather.

Lester Brown, of the Worldwatch Institute, a leading environmental think tank in Washington, D.C., pointed out as far back as 1977
that "deliberate attempts to alter the climate are becoming increasingly common," raising the prospect of "meteorological warfare as
countries that are hard-pressed to expand food supplies begin to compete for available rainfall." It has turned out to be extremely difficult,
so far, to produce even small-scale changes in weather. But that has
not stopped speculation about very large-scale changes, indeed. Dis-
cussions of global warming are conjuring up horrific images of rising
shorelines around the world as polar icecaps melt. But few today re-
member the breathtaking plan to thaw the Arctic Ocean that was
reputedly put forward by Lenin shortly after the Russian Revolution.

Russia's historic strategic problem was the lack of a warm-water
port for its navy. It has an immense shoreline, but most of it is in the
Siberian north. The waters are icebound, the land frozen. The Arctic
Ocean, however, is fed by freshwater rivers flowing into it from
Siberia. Lenin's plan was to dam these rivers and divert them to the
south. This would unleash tremendous amounts of hydroelectricity
for industrial development; it would warm the Siberian climate, thus
increasing arable land; it would reduce the flow of fresh water into the
ocean, presumably altering its salt concentration and causing ice to
melt; in turn, this would open ports for the Russian navy, giving it
easy access to the rest of the world's seas.

While nothing came of this ecologically terrifying plan, the Soviet

Union as late as 1956 reportedly proposed a joint project with the United States to build a barrier across the Bering Strait that would, as in Lenin's plan, warm the Arctic Ocean. Atomic pumps would speed the water northward, not only benefiting Russia's coastline, but Alaska's as well.

The United States is said to have rejected the plan after Pentagon experts pointed out that it would have inundated the West Coast of America, pushing water levels up an estimated five feet all the way from southern California to Japan.

Undeterred, the Soviets are said to have made a similar proposal to the Japanese, this time to warm the Sea of Okhotsk. All these plans were to confer important strategic advantages for the Russian navy's ships and submarines.

International agreement prohibits "military or other hostile use of environmental modification techniques having widespread, long-lasting or severe effects." But it is unlikely that Saddam Hussein sat up nights reading that clause in the Geneva Disarmament Conference the night before he dumped oil into the Persian Gulf or when he darkened Kuwait's sky with a petroleum cloud.

The revolutionary technologies of tomorrow, unless anticipated and rechanneled, open new vistas of destruction for the planet. A new Third Wave war-form is emerging. Does anyone seriously think that yesterday's anti-war approaches are still adequate?

In 1975, at a hearing on the future of the United Nations before the Committee on Foreign Relations of the U.S. Senate, the late author and anti-nuke campaigner Norman Cousins was asked what should be done to prevent the further proliferation of nuclear weapons. Verging on despair, he said the world should have thought about that thirty years earlier.

When it came our turn to testify, we suggested to the senators that they and the world should start worrying about the weapons of thirty years hence. The same holds true today. Myopia and lack of imagination are diseases that afflict warriors and anti-warriors alike.

15

WAR WITHOUT BLOOD?

THE WORLD MEDIA discovered so-called "smart" weapons decades after their first use, and long after General Morelli began explaining their significance to us. The media have not yet discovered an entirely new class of weapons that could, in time, hold even greater significance — weapons designed to keep people alive.

We are at the point in history — the last half century, say — when the maximization of lethality has reached its outer limits: the point at which nuclear arms could, at least in theory, threaten the very existence of the planet . . . when the push for added lethality in a weapon of mass destruction defeated itself . . . when both nuclear superpowers actually concluded that their strategic weapons were, if anything, too lethal. It is, in fact, the point of dialectical negation, the moment when history begins to reverse itself.

Today a new arms race may be about to dawn on the planet — a push for weapons that minimize, rather than maximize, lethality. If so, the world will owe a debt to an unusual wife-and-husband team that has been quietly toiling away for years to take much of the blood out of warfare.

In May 1993 U.S. attorney general Janet Reno appeared before the U.S. Congress to describe the role played by the Federal Bureau of Investigation in its apocalyptic standoff with a cult in Waco, Texas. The fire that swept the Branch Davidian cult compound took seventy-two lives and triggered recriminations on all sides. Reno told the members of Congress that, during the deliberations that led to the

FBI assault, she wished that there were some "magic" non-lethal weapon that could have been used to save lives, especially those of the children held by the cult.

Someday, thanks in part to Janet Morris and her husband, Chris, there will be.

Tough-minded and tough-talking, Janet and Chris Morris are not experts on policing. They focus on military matters. They begin with no illusions about the morality or trustworthiness of nation-states. They won't be found among peace picketers carrying signs deploring war. Instead, until recently, one found them in the basement of the Pentagon or in the offices of the U.S. Global Strategy Council in Washington. The GSC is a private organization headed by Ray Cline, a bearded, graying bear of a man who used to be a deputy director of the CIA. In an earlier incarnation, in 1950, Cline contributed to the famous National Security Council Memorandum-68, which first spelled out the containment of Soviet communism as a formal U.S. policy.

When Janet Morris and her husband decided to devote years of their lives to taking bloodshed out of battle, they went to Cline, a family friend. He brought them into the Council and helped them line up a group of hard-nosed advisers that included Maj. Gen. Christopher Adams, the former commander of the Strategic Air Command, Gen. Edward Meyer, the former army chief of staff, and scientist Lowell Wood of the Lawrence Livermore National Laboratory. With this much braid, brass, and brains behind them, the Morrises set to work. They became, at least for a time, the world's most impassioned, articulate advocates of non-lethality.

Janet Morris is forty-seven, an intense woman with gray-streaked hair that falls down her back to her waist. On the hot summer day when we met, she wore black boots, gray slacks, a light plaid jacket, and aviator sunglasses. Impatient with small talk, she thinks and talks at electronic speeds. Chris, raised a Quaker, is a former musician who found his way into computers. Soft-spoken and slightly balding, he now wears his hair in a fashionable ponytail. The Morrises form a tight intellectual team.

Reflecting the shift away from theories of mass destruction, today's military men are fond of repeating Sun-tzu's famous lines, "To win one hundred victories in one hundred battles is not the acme of skill. To subdue the enemy without fighting is the acme of skill." Janet and Chris Morris push this insight to a new level of strategic theory.

In a nutshell, they argue that a host of new technologies exist, or soon could, that might be used to defeat an enemy — and not just a suicide cult — with absolutely minimal bloodshed. These non-lethal technologies, however, are scattered, unintegrated, and outside the military frame of reference, with its traditional emphasis on killing the enemy. What is needed, they believe, is a complete reconceptualization of war and diplomacy alike. Their mission has been to develop a strategy and doctrine for non-deadly war.

They define as "non-lethal" those technologies "which can anticipate, detect, preclude, or negate the use of lethal means, thereby minimizing the killing of people."

The Morrises began by putting together a lengthy list of militarily useful technologies that meet their criteria for non-lethality. To make their list, a technology must be "fiscally responsible, life conserving, and environmentally friendly." It must not have as its primary purpose "the taking of human life."

It must not be pie-in-the-sky. It has to "offer something right out of the box . . . that will not be expensive." Their list, they claim, excludes "$800,000,000 research projects that take twenty years and may or may not prove out within the lifetime of the researcher." While some believe they are overoptimistic, the Morrises claim a vast arsenal of non-lethal weapons could, in fact, be ready within five years. Their reports for the GSC describe the technologies on their list as ready to go, mature, or needing little more than five years to develop.

Finally, they have also excluded from their list chemical, biological, or other weapons whose use is restricted by international law, treaty, or convention.

ULTRASECRET LABS

The Morrises are openly suspicious of some of the work being done in ultrasecret military labs under the banner of non-lethality but which can create what Janet Morris terms "perverted versions of non-lethal weapons . . . charming little things like two-stage [weapons] with [one] stage which only makes a roomful of people sick, but the second stage of which will kill anybody who has been exposed to the first." We especially "need to watch out for extreme chemicals and biologicals," she says. Non-lethal, one supposes, should mean non-lethal.

The Morrises are not woolly-minded about all this. "War," they write, "can never be made humane, clean, or easy. War will always be terrible." Nevertheless, they continue, "a world power deserving of its reputation for humane action should pioneer the principles of non-lethal defense. . . . Technology now allows us the option of stopping aggression while keeping even the enemy alive. We," they tell U.S. policymakers, "must be first among nations to develop the capability."

Given the deep implications of non-lethal weaponry, it is hardly surprising that opinion in the military is split. Says ex–army chief of staff Edward Meyer, a member of the GSC advisory group, "There is a group in the army very strongly for it, and a group very strongly against it." For some, war is killing by definition, and non-lethality is less than "manly."

But that conviction is a relic of yesterday's war-forms, out of synch with the emerging ethics and technology that underpin the Third Wave war-form. The new spirit is evidenced in the words of Perry Smith, the CNN military analyst during the Gulf War, who was once the U.S. Air Force's deputy chief of long-range planning. Says Smith, "Military planners must look beyond the use of bombs and missiles to precisely attack targets. Technology may soon allow the destruction of key elements of a military target without killing soldiers or totally destroying the target. If an enemy tank can be rendered ineffective by preventing the engine from operating or by ruining the gun-firing computers, winning wars through means that are largely non-lethal may be possible."

He is echoed by Col. John Warden, whose air power theories heavily influenced the American strategy in Iraq. Warden sees the Persian Gulf conflict as an historic turning point. It marked, he says, a major change "from the old concept of slaughter into a transition period where we can get the job done much more effectively and at a much lower cost to human lives, to our environment, and even to our budget."

A year after the Gulf War ended the Defense Department officially endorsed the idea of developing technologies and doctrine for systematic non-lethal war — "soft-kill," as it is sometimes called. As interest has risen, the U.S. Naval War College has played at least two formal war games involving non-lethal conflict.

Ironically, the recent U.S. stampede to cut military spending has temporarily anesthetized the initiative, but the very drive toward smaller budgets will encourage the search for cheaper, more precisely selective — and less lethal — forms of combat.

THE INVISIBLE WALL

To appreciate the possibilities of non-lethal weaponry, once systematically developed, we need to imagine some of the situations in which it might be deployed. One can imagine, for example, an attack on Western embassies by an enraged crowd of Islamic extremists in, say, Khartoum, the capital of the Sudan. Mobs ransack a number of embassies but, oddly, despite chants of "Death to America," the American embassy itself is untouched and no U.S. hostages are seized.

As thousands of rioters approach the walled U.S. compound, their leaders fall to the ground vomiting and defecating. Hundreds of protesters double over and appear disoriented. None comes closer to the wall than half a city block away. As the number of nauseated and diarrhea-disabled protesters grows, the crowd breaks and gradually departs, some of its members crying out that Allah is punishing them.

An American embassy spokesperson in Khartoum calls the attack on the other embassies "a barbaric crime against the international community." He refuses to answer questions as to whether the U.S. State Department has recently installed a new "secret weapon" to protect its embassies.

It is known, however, that advanced infrasound generators designed for crowd control have been tested by France and other nations. The devices emit very low-frequency sound waves that can be tuned to cause disorientation, nausea, and loss of bowel control. The effects have been found to be temporary, terminating when the generator is switched off. No permanent aftereffects are known.

American motorists today can mount a small device on their cars to keep deer from running in front of their wheels. Infrasound deterrence operates on the same principle as these deer-savers, and extensions of technologies like these are even more dramatic.

For example, Special Forces troops dropped by parachute or helicopter might be able to wade directly into a hostage-holding mob without fear and without harming anyone. Says Janet Morris, "We think we have identified some interesting countermeasures that might allow our soldier to turn a field on, penetrate the field without harm, pull an incapacitated perpetrator or hostage out of a group of people . . . and move on out."

It is even conceivable, the Morrises say, that protective devices will be built directly into the physical structure of an embassy, turning the

entire building into a kind of transducer that can be tuned to create a
defensive electronic shield as needed.

In a world of raging religious, racial, and regional hostilities, in
which lethal weapons may well be counterproductive, intensifying
hatreds and violence, rather than quelling them, non-lethal weapons
are likely to find growing acceptance.

No one can be sure. But faced with a dilemma like the one in Waco
in the future, it is at least conceivable that the FBI will be able to erect
a set of disabling sound generators around the cult compound and
prevent self-immolation.

Morris cites the Temple Mount massacre in Jerusalem in 1990 as an
example of bloodshed that could have been avoided had an infrasound generator been used to break up the Palestinian crowd that
hurled rocks and chains and iron bars down on Israelis near the Wail-
ing Wall. "If they vomited or defecated on themselves or if they had a
headache," Morris says, "then that's better than anybody being dead."
In the absence of the proposed technologies, twenty-one people died.
Similar examples might be multiplied, from Tienanmen Square to
Timor.

Echoing these ideas, William J. Taylor, Jr., of the Center for Strategic and International Studies, in Washington, D.C., points to the
Balkan and Somalian conflicts as perfect examples of the need
to speed non-lethal development. "Think," he writes, "of what it
would mean if the world community could field forces to separate
and disarm . . . warring factions instead of killing them. Think of
what it would mean if UN peacekeepers had options beyond rubber
bullets or tear gas." In Waco, he notes, the U.S. government used
"technology dating back to 1928 and the result was an inferno of
retribution."

DROWSY DRUG LORDS

Or imagine a raid on the home of a leader of Kurdish heroin traffickers moving drugs from the Bekaa Valley in Lebanon through Turkey into Bulgaria for European distribution. Once tipped off, a
properly armed and trained special forces team of the Turkish army
could use laser rifles to temporarily blind the posted guards, then
spray "calmative" agents into the barracks and bedrooms, and round
up the groggy drug lords and their followers.

Laser rifles are no fantasy. They can damage enemy optical and
infrared equipment. Used against people, they can flash-blind them

temporarily. They can also do permanent harm, depending on the power used and whether the targeted person is using optical equipment like night-vision goggles, which might amplify the light. According to Leonard H. Perroots, a retired director of the U.S. Defense Intelligence Agency, "These devices are advertised openly for sale to worldwide military forces." Tens of thousands are in circulation. Some were used by Soviet troops in Afghanistan against mujahedeen guerrillas.

Similarly, sleep-inducing agents are not just to be found in James Bond movies. A Global Strategy Council listing of non-lethal technical options refers to "calmative agents" as a class. It explains that "when we must incapacitate people as well as equipment, calmatives or sleep agents mixed with DMSO (which quickly delivers chemicals through the skin into the bloodstream) can curb violence and limit casualties wherever full [nuclear, biological, chemical] gear is not worn. In anti-terrorist actions, counterinsurgency, ethnic violence, riot control, or even in select hostage situations, calmative agents offer an underrated tactic whose effectiveness depends only on modern precision and area delivery systems."

All the non-deadly technologies described so far target human beings. But other non-lethal technologies are aimed at an enemy's hardware and software. It doesn't matter how many tanks or planes it has, or how good its radar systems are, if they can't be used where and when it needs them. In fact, the more materiel the enemy has and the more it has spent on it, the worse off it is if that material is put out of commission even temporarily. Thus a key concept in non-lethality theory is "denial of service."

Take, for instance, the concept of "anti-traction." As one GSC document puts it, "Anti-traction makes surfaces slippery. Using airborne delivery systems or human agents, we can spread or spray Teflon-type, environmentally neutral lubricants on railroad tracks, grades, ramps, runways, even stairs and equipment, denying their use for a substantial period." Alternatively, it is also possible to paste things down so they cannot move. "Polymer adhesives, delivered by air or selectively, on the ground, can 'glue' equipment in place and keep it from operating."

Engines can be stalled or stopped. Thus tanks, armored personnel carriers, and trucks can be paralyzed by special munitions that will temporarily "contaminate fuel or change its viscosity to degrade engine function." Directed energy weapons could change the molecular structure of their targets, keeping planes on the ground.

Then there is "liquid metal embrittlement." It is possible to wage a kind of "graffiti" war by using a felt marker or a spray can to apply a colorless chemical to key components of metal structures like bridge pylons, airport facilities, elevators, or weapons. The fluid causes them to become brittle, breakable, and hence unusable.

Later on we shall see that the concept of "denial of service" through non-lethal means has vastly larger potentials than this short list suggests. For now, however, it is enough to recognize the growing significance of non-lethality in general. There are, no doubt, plenty of grounds for heated controversy about the cost and technical feasibility of non-lethal weaponry. But it is no longer possible to dismiss the fact that new, Third Wave technologies can be designed to minimize casualties on all sides. We may not be able to take war out of the future, but it appears likely that we can take some of the blood out of war.

Not even Chris and Janet Morris believe war can be made truly bloodless. In any armed conflict, someone is going to get hurt. As she puts it, "You're going to have incidental, accidental and corollary casualties, as you will with anything else that's heavy enough if you drop it on somebody's head. We're not guaranteeing a bloodless environment."

Nor will non-lethal weapons replace lethal arms in any foreseeable future. "We're not suggesting non-lethal units, suicide commando squads or anything of the sort. It's not a substitute, at this time . . . for a conventional force where the lives of our soldiers are at risk." Nevertheless, the very range of new technologies available — from computer viruses to "calmatives" — makes it possible to marshal them in a systematic way that could amplify their effect and reduce reliance on lethal means.

Bit by bit, non-lethality is creeping into doctrinal thinking. But it is a long hard slog against entrenched attitudes. In September 1992, after a year of internal debate, the U.S. Army issued a draft paper called "Operations Concept for Disabling Measures." It was intended to minimize large-scale casualties in populations caught in a war zone, as well as damage to the environment and the infrastructure. The document announced expanded research under the army's "Low Collateral Damage Munitions" program. But almost no attention is paid to non-lethality in a June 1993 revision of the official doctrine. So it is clear that the concept remains controversial.

What needs underscoring, however, is that non-lethality and the new doctrines emerging from the military are both products of Third Wave societies whose economic lifeblood is information, electronics, computers, communication, and mediatization — the rising ubiquity and importance of the media.

THE POLITICS OF NON-LETHALITY

As with many other Third Wave phenomena, from interactive television to genetic engineering, non-lethal technologies bring risks and moral perplexities as well as humanitarian reward.

To begin with, it should be amply apparent by now that many of these weapons, if wielded by terrorists or criminals, rather than by the "good guys," could serve as a force multiplier for them. On a small scale, what might terrorists or irresponsible political protesters be able to do to highly vulnerable structures in a city, airport, or dam with a felt pen or a spray can containing an "embrittlement" agent? Imagine today's graffiti "taggers" with chemicals like these in their spray cans. It is fine to speak of immobilizing tanks with anti-traction. But what might urban guerrillas do to police cars parked outside the local station house? And if sociopathic computer hackers infect computers with viruses, what might they, or others, someday do with microwave weaponry?

Even when used by legitimate authorities, non-lethal weapons raise profound political and moral questions. Janet Reno might have been able to subdue the Koresh cult in Waco without significant violence, and thereby save at least some of the children who died.

But many of these weapons can be used by repressive states against their own peacefully protesting citizens. Some of the technologies are so suited to use for crowd control or protest-busting that democracies may have to write new rules of engagement for their police.

Then there is the question of how to categorize weapons. What weapons are truly non-lethal? Some have "adjustable lethality" — operated at low power, they cause minimum temporary damage; tune them up, and they can kill. Are they non-lethal or not? To the credit of the Morrises and the Global Strategy Council they have not brushed these and other problems aside in blind enthusiasm for non-lethality.

It is precisely because they recognize the risks — and especially the risks for democracy — that the Morrises wanted to pull back the nearly impenetrable cloak of secrecy placed on this breed of weaponry by the so-called "black acquisitions people" in the ultrasecret labs and services. So tight is this concealment that the Morrises themselves, both of whom hold high security clearances, have been denied access to some of the ongoing work.

Chris and Janet Morris admit the need for a degree of military secrecy, but they argue forcefully that non-lethal warfare is so important a part of the future it must be opened to wider public debate and discussion. They have riled some Department of Defense officials by arguing in favor of bringing the development of non-lethal weaponry under congressional scrutiny. There are, they say, dangerous human rights issues involved that should not be left for the military to decide by default.

Similarly, the wider introduction of non-lethal warfare methods raises new questions at the geopolitical level. If, for example, the United States — today the world's only superpower — relied more on non-lethal methods and less on conventional force, would other nations mistake this restraint for weakness? Would the rise of non-lethal weaponry encourage adventurism or, alternatively, lead to false expectations of unilateral disarmament? Or both?

Might it lead to a new rivalry — a race by countries to spread non-lethal arms everywhere? Might that ultimately lead to less killing — and less democracy as well — if states can blind, dazzle, disorient, and otherwise defeat their critics non-lethally? And if there were to be a non-lethal arms race, which nations would stand to gain the most? Which are most capable of producing the sophisticated new type of arms? Will non-lethality open a vast new field for Japanese technology? Today Article Nine of Japan's constitution still prohibits the export of arms. But what is the definition of arms? And do non-lethal devices fall within its scope?

WHEN DIPLOMATS FAIL...

In the past, when diplomats fell silent, guns very often began to boom. Tomorrow, according to the U.S. Global Strategy Council, if diplomatic talks fail, governments may be able to apply non-lethal measures before engaging in traditional, bloody war.

Janet Morris believes that this "area between when diplomacy fails and the first shot is fired is an area that has never been quantifiable before. It has been a non-space." Non-lethality thus emerges not as a simple replacement for war or an extension of peace but as something different — something radically new in global affairs: an intermediate phenomenon, a pausing place, an arena for contest in which more outcomes could be decided bloodlessly. It is a revolutionary form of military action that faithfully reflects the emerging Third Wave civilization.

But it raises as many questions for anti-war as it does for war. Can one formulate not merely a war doctrine for non-lethality but an anti-war doctrine as well? That question should stimulate fresh thinking among politicians, defense contractors, armies, diplomats, and peace movements around the earth as we race into a period of ethnic and tribal upheavals, secessionist movements, civil wars, and insurrections — the bloody birth pangs of tomorrow's world.

What is becoming apparent now is that the military revolution that began with AirLand Battle and made its first public appearance during the Gulf War is still only in its infancy. The years ahead, despite budget cuts and rhetoric about peace in the world, will see military doctrines around the world change in response to new challenges and new technologies. In a world of niche wars, niche warriors can be expected to flourish. In a world that is becoming ever more dependent on space for communications, weather reports, and myriad other things, the military dependence on space will grow. In a world whose factories are becoming ever more computerized and automated, war, too, can be expected to rely on computers and automation, including robotization. As new technical triumphs erupt from the world's laboratories, armies, for good or for ill, will seek advantage in everything from genetics to nano-technology, fulfilling and exceeding even the wildest dreams of today's Da Vinci–like dreamers. At the same time, in a world in which the slaughter of civilians sometimes has counterproductive political consequences, the development of non-lethal weaponry will proceed apace. Combining weapons of high selectivity with others with non-lethal effects points, more hopefully, toward a possible reduction in indiscriminate death.

Each of these developments will be incorporated into the still-embryonic Third Wave war-form that reflects the still-embryonic Third Wave economy and civilization of the future. But it would be a serious mistake to think that the dominant war-form of tomorrow will be exclusively defined by things like satellites, robotry, or non-lethal weaponry. For the common element that binds all these elements together is not hardware — not tanks or planes or missiles, not satellites or nano-weapons or laser rifles. The common thread is intangible. It is the same resource that defines the emergent system for wealth creation and the society of tomorrow: knowledge.

Thus we begin to see a clear progression. The Third Wave war-form began with AirLand Battle. The Gulf War offered only a pale hint as to the further development of the new war-form. In the decades to come it will be broadened to incorporate new possibilities provided by advancing technology. But even these do not, and cannot, complete its development.

For the evolution of the Third Wave war-form will not be complete until its central resource is understood and deployed. Thus the final development of Third Wave war may well be the conscious design of something the world has not yet seen: competitive knowledge strategies.

With that, war moves to a totally new level.

PART FOUR

KNOWLEDGE

16

THE KNOWLEDGE WARRIORS

As THE THIRD WAVE WAR-FORM takes shape, a new breed of "knowledge warriors" has begun to emerge — intellectuals in and out of uniform dedicated to the idea that knowledge can win, or prevent, wars. If we look at what they are doing, we discover a step-by-step progression from initially narrow technical concerns toward a sweeping conception of what will someday be called "knowledge strategy."

Paul Strassmann is a brilliant, intense, Czech-born information scientist. Formerly a strategic planner and head of information services for the Xerox Corporation, he is the author of important studies of the relationship between computers, worker productivity, and corporate profitability in the civilian economy. More recently he served as Director of Defense Information in the Pentagon — the Chief Information Officer of the American military.

Strassmann is a walking data bank of information technology — types of computers, software, networks, telecommunications protocols, and much else besides. But more than a narrow technologist, he has thought a lot about the economics of information. He brings, in addition, a rare historical sweep to his work. (As a sideline during his Xerox years, Strassmann and his wife, Mona, jointly created an elegant museum devoted to the history of communication, from the invention of writing to the computer.) His personal history, too, has shaped his ideas about warfare. As a boy in World War II, he fought against the Nazis with a Czech guerrilla commando group.

"The history of warfare," says Strassmann, "is the history of doctrine. . . . We have a doctrine for landing on beaches, a doctrine for bombing, a doctrine for AirLand Battle. . . . What is missing . . . is doctrine for information."

It may not be missing for long. In February 1993 West Point, the U.S. Army academy, appointed Strassmann a Visiting Professor of Information Management. Simultaneously, the National Defense University at Fort McNair, in Washington, introduced the first course on Information Warfare.

The NDU and West Point are not alone. In the office of the U.S. Secretary of Defense there is a unit called "Net Assessment" whose primary task is weighing the relative strength of opposing military forces. Headed by Andy Marshall, this unit has shown a strong interest in information warfare and what might be called info-doctrine. Outside the Pentagon, a private think tank called TASC, the Analytic Sciences Corporation, is also gearing up for work on the issue. Other armies, too, in response to the Gulf War, are thinking about information doctrine, if only in terms of defense against an informationally superior America.

So far much of this doctrinal discussion still focuses on the details of electronic warfare — knocking out an adversary's radar, infecting his computers with viruses, using missiles to destroy his command and intelligence centers, "spoofing" his equipment by sending false signals, and using other means to deceive him. But Strassmann, Marshall, and the other military intellectuals are thinking beyond practical how-to doctrine to the broader realm of high-level strategy as well.

Duane Andrews is Strassmann's old boss in the Pentagon. Andrews, who served as Assistant Secretary of Defense for C³I ("Command, Control, Communications and Intelligence"), underlined the difference when he termed information a "strategic asset." That means it is not just a matter of battlefield intelligence or tactical attacks on the other side's radar or telephone networks, but a powerful lever capable of altering high-level decisions by the opponent. More recently, Andrews spoke of "knowledge warfare" in which "each side will try to shape enemy actions by manipulating the flow of intelligence and information."

A more formal description is to be found in a jargon-studded document released on May 6, 1993, by the office of the U.S. Joint Chiefs of Staff. This "Memorandum of Policy No. 30" defines "command and control" (abbreviated as C²), as the system by which authority and direction are exercised by legitimate commanders.

It defines command and control warfare as the "integrated use of operations security ... military deception, psychological operations ... electronic warfare ... and physical destruction, mutually supported by intelligence, to deny information to, influence, degrade, or destroy adversary C^2 capabilities, while protecting friendly C^2 capabilities against such actions." Properly executed, the report declares, command and control warfare "offers the commander the potential to deliver a KNOCKOUT PUNCH before the outbreak of traditional hostilities."

The memo widens the official parameters around the concept of information warfare by placing more emphasis on intelligence and by extending the scope to include psychological operations aimed at influencing the "emotions, motives, objective reasoning, and ultimately the behavior" of others.

As an official statement of Pentagon policy, the document is necessarily filled with carefully hedged language, legalistic definitions, and specific instructions and assignments. The intellectual discussion of information warfare in the defense community, however, goes well beyond these limits.

Thus a far broader, theoretical "take" on the subject is found in the work of two scholars at the RAND Corporation in Santa Monica, California, David Ronfeldt and John Arquilla. In a preliminary overview of what they call "cyberwar," they touch on broad strategic questions. Arquilla, clean-cut and soft-spoken, served as a consultant to General Schwarzkopf's Central Command throughout the Gulf War. Ronfeldt is a bearded, tweedy, even more soft-spoken social scientist who has studied the political and military effects of the computer revolution.

Cyberwar, for them, implies "trying to know all about an adversary while keeping it from knowing much about oneself. It means turning the 'balance of information and knowledge' in one's favor, especially if the balance of forces is not." And exactly as in the civilian economy, it means "using knowledge so that less capital and labor may have to be expended."

The gabble of terminology — Info-Doctrine, Cyberwar, C^2 Warfare, and other terms mercifully omitted here — reflects the still-primitive stage of discussion. No one has yet taken what appears to be the final step in this progression — the formulation of a systematic, capstone concept of military "knowledge strategy."

Certain things are nevertheless clear. Any military — like any company or corporation — has to perform at least four key functions with respect to knowledge. It must acquire, process, distribute, and protect information, while selectively denying or distributing it to its adversaries and/or allies. If, therefore, we break each of these down into their components we can begin to construct a comprehensive framework for knowledge strategy — a key to many, if not most, of tomorrow's military victories.

SILICON VALLEY'S SECRET

Take acquisition — producing or purchasing the knowledge needed by the military.

Armies, like everyone else, acquire information in myriad ways — from the media, from research and development, from intelligence, from the culture at large and other sources. A systematic acquisition strategy would list these and determine which ones need to be improved.

For example, America's clear technological edge in warfare sprang, in good measure, from the fact that the Defense Department spends nearly $40 billion annually to conduct or contract out defense-related R & D.

During the Second Wave era, military technology in the United States advanced at lightning speed and spun off innovation after innovation into the civilian economy. Today a role reversal has occurred. In the fast-paced Third Wave economy, technical breakthroughs come faster in the civilian sector and spin off into the defense industries. This calls for a strategic reexamination of R & D priorities and a restructuring of relations between the military and civilian science and technology.

An alternative way to obtain valuable knowledge is, of course, through espionage and intelligence activities. Intelligence is obviously central to any conception of knowledge-based warfare. The coming upheaval in intelligence is so profound, however, as to warrant fuller treatment than can be given here. (See Chapter 17, "The Future of the Spy.")

Finally, acquisition can also involve things like organized, strategic brain drains. During World War II there was a lively (sometimes deadly) competition for scientific brainpower. The Nazis severely damaged their own military effectiveness by driving out or exterminating some of the best scientific minds of Europe, many of them Jewish. The Allies sought out these minds and put them to work on the Manhattan Project, which produced the first A-bomb. Others took prominent roles in fields ranging from strategic studies and political science to psychoanalysis. Conversely, the Allies tried to kidnap German atomic scientists to prevent Hitler from acquiring his own nuclear bomb.

The military and commercial significance of such positive and negative brain drains is likely to grow as information and know-how diffuse around the world. To quote the influential management theorist Tom Peters, "One of Silicon Valley's great secrets is stealing human capital from the Third World. Maybe the natives [of the Valley] are leaving. That is more than made up for by Indians and Taiwanese coming in."

Thus tomorrow's military knowledge strategists may well design sophisticated, long-range policies to suck certain kinds of brainpower out of target countries and transfer it to their own. Alternatively, knowledge strategies will increasingly include plans designed to discourage or deny the movement of key scientists or engineers to potential adversaries. Recent efforts to keep Russian scientists from emigrating to Iran and North Korea are only the latest round in a game that will be played for enormous strategic stakes.

Clever knowledge strategists will pay as much attention to "knowledge procurement" tomorrow as is paid today to the procurement of hardware.

THE SOFTWARE SOLDIERS

Advanced armies, like companies, also have to store and process information in huge quantities. Increasingly, as we know, that requires immense investments in information technology, or I-T.

Military I-T includes computer systems of every conceivable size and type. The nature, distribution, capacity, usability, and flexibility of such systems, including their links with radar, air defenses, and satellite and communications networks, will distinguish advanced armies from one another.

In the United States, much of the work done by Duane Andrews and Paul Strassmann and by their respective successors in the Pentagon, Charles A. Hawkins, Jr., and Cynthia Kendall, involved trying to rationalize, upgrade, and improve these vast systems. Hawkins, an engineer, rose through military intelligence. Kendall, the Deputy Assistant Secretary of Defense for Information Systems, was trained in mathematics and operations research. She joined the Department of Defense in 1970.

More important than the actual hardware that Hawkins and Kendall oversee is the constantly changing inventory of software on which it depends. In the Gulf War television cameras, ravenous for dramatic visuals, focused on F-14 Tomcat fighters roaring off the decks of carriers, Apache helicopters swooping over the desert, M1A1 Abrams tanks growling over the sands, and Tomahawk missiles singling out their targets. Pieces of hardware became overnight "stars." But the real "star" was the invisible software that processed, analyzed, and distributed data, though no television watcher ever saw those who produced and maintained it — America's software soldiers. Most were civilians.

Software is changing military balances in the world. Today weapons systems are mounted on or delivered by what the jargon calls "platforms." A platform can be a missile, a plane, a ship, or even a truck. And what the military is learning is that cheap, low-tech platforms operated by poor, small nations can now deliver high-tech smart firepower — if the weapons themselves are equipped with smart software. Stupid bombs can often have their I.Q. raised by the addition of retrofitted components dependent on software for their manufacture or operation.

In the Second Wave era, military spies paid special attention to an adversary's machine tools because they were needed to make other tools needed for producing arms. Today the "machine tool" that counts most is the software used to manufacture software that manufactures software that manufactures software. For much of the processing of data into practical information and knowledge is dependent on it. The sophistication, flexibility, and security of the military software base is crucial.

Policies that guide the development and use of information technology in general, and software in particular, are a crucial component of knowledge strategy.

IS UNCLE SAM LISTENING?

Even if it is acquired and suitably processed, knowledge is useless in the wrong hands or heads at the wrong time. Hence the military's need for various ways of distributing it as needed.

"The services," says Lt. Gen. James S. Cassity, "put more electronics communication connectivity into the Gulf in ninety days than we put in Europe in forty years." Connectivity is the jargon for networks, and the kind of networks that are constructed, and who is admitted to them, is closely tied into high-level strategic concerns.

Ambitious plans are afoot, for example, to create a single seamless, globe-girdling military communications network that goes beyond the U.S. forces — a modular system that can be shared by the forces of many nations at once. Just as more and more businesses are integrating operations globally, forming consortia, and linking their computer systems and communications networks to those of their corporate allies, so, too, is the military — on a far larger scale. The problem with alliances — commercial and military alike — is that coordination is extremely difficult.

Even among NATO nations in Europe, even after four decades of cooperation, battlefield management systems cannot yet communicate tactical information to one another because of incompatibility. Although NATO laid down common standards, neither the British Ptarmigan system nor the French RITA radios meet the standards. The Tower of Babel problem is even worse elsewhere. After the invasion of Kuwait, it took many weeks to link the military communications systems of Saudi Arabia, Qatar, Oman, Bahrain, and the Emirates with those of the United States.

The envisioned new network is intended to overcome precisely such problems and to make combined operations with allies smoother than in the past. According to Mary Ruscavage, a deputy director of the U.S. Army Communications-Electronics Command at Fort Monmouth, in New Jersey, "We are trying to develop a generic architecture and to take into account all the types of equipment a country has."

The nature of communications networks presupposes often unspoken strategic assumptions. In this case, the notion of a shared global network into which other nations can plug clearly reflects the American strategic assumption that it will, in the future, fight in combination with allies, rather than as a lone "world cop."

The proposed system conjures up images of a future marked by temporary, plug-in/plug-out alliances — in keeping with the fluidity of conditions in the post–Cold War world. It could simplify future United Nations operations.

But it also raises the question of whether, if the United States basically designs the system, it becomes possible for America to read all the messages flowing through the network. (Not necessarily, it is argued, because individual nations could specify their own "crypto," as coding is known. But suspicions still flourish.)

Stuart Slade, a London-based information scientist and military analyst for Forecast International, points to another, deeper political implication of the new command, control, and communications systems. Not every army in the world is culturally or politically (let alone technologically) capable of using them. "These systems," he explains, "depend on one thing — and that is the ability to exchange information, to swap data, and to promote a free flow of information around the network, so that people can assemble their tactical pictures, they can relate their stuff together. What we have actually got is a 'politically correct' weapon system.

"Societies that freeze the flow of communications, the free flow of ideas and data, will not, by definition, be able to make much use of such systems. . . . The Iraqi system is a tree. You've got Saddam Hussein at the top. If you break that kind of system at any point, it can be catastrophic, especially if the division commander, severed from the top of the tree, knows that his reward for using his initiative is a .357 [bullet] delivered to the back of his head."

Since advanced networks permit users to communicate among themselves at all levels of the hierarchy, it means that captains can talk to other captains, colonels to other colonels, without the messages first going to the top of the pyramid. But that is precisely what totalitarian presidents and prime ministers may not want.

"There are quite a few countries," Slade suggests, including China, that would find such a system politically dangerous. "There are," he says, "countries in Africa, for example, where, if you gave battalion commanders the ability to talk to one another, without someone standing over their heads, within six months one battalion commander would be President and the other the Minister of Defense."

This is why, he believes, the new communications networks favor democratic nations.

DE-LEARNING AND RE-LEARNING

Crucial as it is, however, communication is only one part of the knowledge-distribution system of the armed forces. Third Wave militaries place a massive emphasis on training and education at every level, and their systems for delivering the right training to the right person are part of the knowledge-distribution process.

As in business, learning, de-learning, and re-learning has become a continuous process in every occupational category in the military. Training organizations are rising in the power pecking order within the various military services. In all branches advanced technologies are being developed to speed learning. Among these, computer-based simulation plays a greater and greater role. For example, using actual video of a key Gulf War engagement, all moves by the tanks of both sides have been fed into a computer, permitting crews to re-fight the battle under varying simulated conditions. One can imagine the day when computer-based training methods and technologies themselves become so valuable that armies try to steal them from one another. Third Wave generals understand that the army that trains the best, learns the fastest, and knows the most has a keen edge that can compensate for many shortfalls. Knowledge is the ultimate substitute for other resources.

Similarly, smart generals understand all too well that wars can be won on the world's television screens as well as on the battlefield.

Among the things that armies distribute are deceptive information, disinformation, propaganda, truth (when it serves them) and powerful media imagery — knowledge along with anti-knowledge.

Propaganda and the media, indeed, will play so politically explosive a role in twenty-first-century knowledge warfare that we devote a later chapter to them (see Chapter 18, "Spin"). Media policy, therefore, along with policies for communication and education, will together comprise the main distribution components of any overall knowledge strategy.

THE SEVERED HAND

But no knowledge strategy is complete without a final, fourth component — the defense of one's own knowledge assets against enemy attack. For the sword of knowledge cuts two ways. It can be used in offense. It can destroy an opponent even before his first lunge. But it can also cut off the very hand that wields it. Right now, the hand that wields it best is American.

No nation in the world is more vulnerable to the loss of its knowledge assets. And no nation has more to lose.

This point is hammered home by Neil Munro, a thirty-one-year-old Dubliner with a faint Irish brogue who transplanted himself to

America in 1984 with a master's degree in war studies under his arm. Today he is one of the best-informed experts on the rise of information-war thinking, from its origins in electronic warfare to the latest Pentagon twists and turns.

Author of *The Quick and the Dead*, a key book about electronic combat, he is a staff writer for *Defense News*, an authoritative weekly that claims 1,315 American generals and admirals among its readers — not to mention another 2,419 high-ranking officers in foreign armies and navies around the world. *DN* is also widely read by defense industry executives, politicians, cabinet ministers, and even, it insists, a few heads of state. In short, when Munro tracks the latest developments in info-war doctrinal thinking, or in software, or in intelligence, his reports land on the desks of the relevant decision makers.

Munro virtually bubbles with adrenaline, his words tumbling out as he talks about information warfare, punctuating his commentary with erudite references to military history. He reflects the intellectual energy forming around the conceptual building blocks that lead toward the ultimate goal of knowledge strategy. But Munro also echoes a persistent warning heard in info-war circles.

Information or knowledge superiority may win wars. But that superiority is exceedingly fragile. "In the past," says Munro, "when you had five thousand tanks and your enemy had only one thousand, you may have had a five-to-one superiority. In information war, you can have a hundred-to-one superiority, but it can all turn on a fuse." Or on a lie. Or on your ability to protect your advantage from those who want to steal it.

The key reason for this fragility is that knowledge, as a resource, differs from all the others. It is inexhaustible. It can be used by both sides simultaneously. And it is nonlinear. That means that small inputs can cause disproportionate consequences. A small bit of the right information can provide an immense strategic or tactical advantage. The denial of a small bit of information can have catastrophic effects.

In the afterglow of the military victory in the Gulf, American attention focused on the ways in which U.S. forces were able to "blind" Saddam Hussein by knocking out his information and communication assets. Since then, concern verging on alarm has been growing in defense circles over the ways in which an enemy might blind the United States, instead.

INFO-TERROR

On January 19, 1991, in the allied air attack on Baghdad, the U.S. Navy used Tomahawk cruise missiles to deliver what *Defense News* described as "a new class of highly secret, non-nuclear electromagnetic pulse warheads" to disrupt or destroy Iraqi electronic systems. Such weapons cause no overt physical damage but can "fry" the components of radar, electronic networks, and computers.

On February 26, 1993, a crude bomb exploded in the World Trade Towers in Manhattan, killing six, injuring over a thousand people, and disrupting the activities of hundreds of businesses close to the financial center of New York.

Imagine what might have occurred if some of Saddam Hussein's nuclear physicists had created for him a crude electromagnetic pulse warhead and, during the Gulf conflict, an "info-terrorist" had delivered it to the World Trade Towers or the Wall Street district. The ensuing financial chaos — with bank transfer networks, stock and bond markets, commodity trading systems, credit card networks, telephone and data transmission lines, Quotron machines, and general commercial communications disrupted or destroyed — would have sent a financial shock wave across the world. Nor does one need such sophisticated weaponry to accomplish a similar effect. Even primitive devices planted at unprotected "knowledge nodes" can create havoc if systems lack adequate hardening, fail-safe mechanisms, or backups.

Says communications consultant Winn Schwartau of Inter-Pact, "With over 100 million computers inextricably tying us all together through the most complex array of land and satellite based communication systems . . . government and commercial computer systems are so poorly protected today that they can be essentially considered defenseless. An electronic Pearl Harbor is waiting to happen."

A report of the U.S. General Accounting Office to Congress voices similar concern. GAO worries that Fedwire, an electronic fund transfer network that handled $253 trillion in money transfers in 1988 alone, suffers from security weaknesses and needs "stringent security provisions." Paul Strassmann, hardly an excitable personality and no sensationalist, warns of "info-terrorist brigades."

Booz Allen & Hamilton, the consulting firm, has conducted a study of communications in New York and found that major financial institutions were operating without any telecommunications backup. Nor are their counterparts in Frankfurt or Paris or Tokyo or London much better off. The report suggested the contrary.

Military systems, while more secure, are hardly impervious. On December 4, 1992, the Pentagon sent a secret message to its commanders in chief in each region ordering them to get busy protecting their electronic networks and computers. It is not just radar and weapons systems that are vulnerable, as we saw earlier, but even things like the computer data bases that contain mobilization plans or lists and locations of spare parts. Said Duane Andrews at the time, "Our information security is atrocious, our operational [secrecy] is atrocious, our communications security is atrocious." As though to underline these harsh words, in June 1993 an "electronic hacker" intercepted calls placed to world leaders by the staff of U.S. secretary of state Warren Christopher. The calls were intended to alert them to the U.S. missile strike against the Iraqi intelligence headquarters in Baghdad.

So many media stories have appeared about computer hackers who illegally penetrate corporate or government computers that it is hardly necessary to repeat them. But misconceptions still abound. While hackers have been smeared with broad-brush attacks for illegally entering or destroying computer systems, most are, in fact, careful not to damage information or to act illegally. Those who do harm are called "crackers" by the hackers.

Whatever the terms, it is now possible for a Hindu fanatic in Hyderabad or a Muslim fanatic in Madras or a deranged nerd in Denver to cause immense damage to people, countries, or, even with some difficulty, to armies 10,000 miles away. Says *Computers in Crisis*, a report of the National Research Council, "Tomorrow's terrorist may be able to do more damage with a keyboard than with a bomb."

Much has been written about computer viruses that can destroy data or swipe both secrets and cash. They can implant false messages, alter records, and engage in espionage, searching for data and transmitting to an adversary. If they can gain access to the appropriate networks, they can, at least in theory, arm, disarm, or retarget weapons.

Early viruses were introduced into public networks and spread indiscriminately from machine to machine. Computer watchers now worry about the so-called "cruise virus" — a smart weapon that is specifically targeted. Its purpose is not to spread indiscriminate damage but to capture a specific password, steal specific information, or destroy a specific hard disk. It is the software equivalent of the intelligent cruise missile.

Once introduced into a network with many computers on it, the virus may lurk or loiter innocently, waiting until an unsuspecting user — a kind of Typhoid Mary carrier — accesses the targeted computer. The virus then hops on board and goes along for the ride. Once inside it launches its destructive payload.

Hans Moravec, in *Mind Children*, describes a defensive weapon he calls a "viral predator" that spreads through a network like an antibody in the immune system, seeking out and killing viruses. But, he notes, "a prey virus can be cosmetically altered so that it is no longer recognizable to a particular predator." Even that, however, does not exhaust the possibilities.

There now exists a program that, in principle, cannot only be planted in a network to replicate itself in thousands of computers, or to cosmetically alter itself according to preprogrammed instructions, but which can be engineered to evolve over time exactly like a biological organism responding to random mutation — an evolutionary virus whose changes are influenced by chance, making it harder for even the most sophisticated virus-killer to find. It is Artificial Life on its way to autonomy.

It is true that advanced Third Wave democracies are more decentralized and have more redundancy built into them than before, and have enormous social and economic resiliency because of this. But there are counterbalancing disabilities. For example, the more advanced and miniaturized the computers and electronics in a system, the less electromagnetic energy is needed to disrupt them. Moreover, Third Wave societies are more open, their work force more mobile, their political and social systems more permissive, and their complacence greater than that of the nations and groups that wish them ill. For these reasons, if no other, any worthwhile military knowledge strategy must address such security issues, along with questions about acquisition, processing, and distribution of knowledge.

In sum, an army's comprehensive knowledge strategy will have to deal with all four of the key functions — acquisition, processing, distribution, and protection. Each of these is, in fact, interrelated. Protection must be extended to all of these knowledge functions. Information systems for processing touch on all of these functions. It is not possible to separate communications from computers. To protect the military knowledge system requires the acquisition of counterintelligence. How these are to be integrated will occupy the knowledge strategists for a long time to come.

Beyond this — and beyond the scope of this book — is an even larger fact of life. Each of these four knowledge functions in the military has a precise civilian analog. The ultimate strength of a Third Wave military rests on the strength of the civil order it serves, which, in turn, increasingly depends on the society's own knowledge strategy.

That means, for better or worse, that the soldier and the civilian are informationally intertwined. How well the civilian world — business, government, nonprofit associations — acquires, processes, distributes, and protects its knowledge assets deeply affects how well the military will carry out its tasks.

The continuing enhancement and defense of these assets are preconditions for the survival of Third Wave societies in the trisected global system of the twenty-first century.

What we already see, therefore, is the progression of military thinking beyond its early conceptions of electronic warfare, beyond the current definitions of "command and control warfare," and even beyond the more general notion of "information warfare."

For decades to come, therefore, many of the best military minds will be assigned to the task of further defining the components of knowledge warfare, identifying their complex interrelationships, and building "knowledge models" that yield strategic options. These will be the womb out of which full-blown knowledge strategies will be born.

For the design of knowledge strategies is the next stage in the further development of the Third Wave war-form — to which, as we'll see, the peace-form of tomorrow will have to respond.

To arrive at an appropriate knowledge strategy, however, each country or military force will face its own unique challenges. For the United States, with the most advanced military in the world, it implies radical restructure of some of its most important and deeply entrenched "national security" organizations of the Second Wave era.

17

THE FUTURE
OF THE SPY

Forty minutes from the Metropole Hotel in Moscow, we approached the nondescript apartment building. We stamped the snow off our shoes and entered. Mailboxes lined one side of the darkened lobby, some open with papers stuffed in them. We took a small elevator up, then found a warm greeting on the landing. Soon we were comfortably seated in Oleg Kalugin's living room. A well-built man in his early fifties, Kalugin speaks perfect English. He smiles and hands you his name card, which identifies him cryptically as "Expert." It gives no hint of the kind of expertise he has.

Oleg Kalugin was the Soviet Union's chief spy in Washington during some of the hottest years of the Cold War. It is a far cry from the days when he "ran" John Anthony Walker, the American naval officer who peddled U.S. codes, from the days when Kalugin sat in the Soviet embassy on Sixteenth Street reading documents stolen from the super-secret National Security Agency, or later, when he would visit with Kim Philby, one of the master spies of the century. Today Kalugin, once the KGB's youngest general, makes appearances on CNN, meets with high officials of the CIA and FBI, and thinks back over his career.

In the course of several hours, we spoke about the possibility, which he regards as unlikely, that some Soviet spies and networks in various countries have shifted allegiance and gone to work for other nations. He gave us his private assessment of the attempted coup that led to the downfall of Gorbachev, and he described his hopes for a peaceful future.

Kalugin has become a vocal critic of intelligence as it was practiced during the Cold War. He is even more critical of what he sees happening today — notably the Russian government's decision to create an "Academy for State Security" in which a new generation will be taught what he describes as "the same old approaches, the same disciplines" as in the days of the KGB. Some of his former colleagues are outraged at his public criticisms of the espionage agency he once served. But Kalugin is a living symbol of the remarkable changes transforming the world espionage industry.

Among all the "national security" institutions, none have a deeper need for restructure and reconceptualization than those devoted to foreign intelligence. Intelligence, as we've seen, is an essential component of any military knowledge strategy. But as the Third Wave warform takes shape, either intelligence itself assumes a Third Wave form, meaning it reflects the new role of information, communication, and knowledge in society, or it becomes costly, irrelevant, or dangerously misleading.

HOOKERS AND SPORTS CARS

Washington currently reverberates with voices crying for drastic reduction or even wholesale dismemberment of America's spy agencies. But, as with defense spending generally, most of the demands for crash cuts reflect short-range political pressures rather than any grand global strategy or reconceptualization of intelligence, as such.

Thus the ever-influential *New York Times* calls for a shutdown of satellites that monitor telephone calls and missile telemetry; praises the fact that the CIA has only nine analysts paying attention to the Russian military (down from 125); and thinks Iran bears watching, but casually announces that the rest of the world is "pretty well covered."

Such offhand confidence seems misplaced when the former Soviet military still controls thousands of both strategic and tactical nuclear weapons, when the country remains potentially explosive, and rogue elements of the old military could still play a revolutionary role in determining the future. Self-imposed deafness seems hardly sensible in a world that is proliferating missiles and warheads at high speed. In terms of potential for triggering global instability, Iran is not the only place that "bears watching." And the "rest of the world" is assuredly not "pretty well covered," as the pages of the *Times* itself reveal.

Since at least the 1970s it was universally assumed that Kim Il

Sung, the Communist dictator of North Korea, was grooming his son, Kim Jong Il, to succeed him in office. But almost nothing has been known about the son, beyond a reported penchant for imported cars and Swedish hookers. In March 1993 the *Times* reported that "the CIA apparently discovered only recently that he has two children, an important fact in a government with a dynastic tradition." That it took so long for Western intelligence to determine so basic a political fact hardly evidences good "coverage."

THE GM PROBLEM

For the United States, foreign intelligence was a $30 billion-a-year enterprise. Its main institutions, the Central Intelligence Agency, the Defense Intelligence Agency, the National Security Agency, and the National Reconnaissance Office, were classical Second Wave organizations. They were huge, bureaucratic, centralized, and highly secretive. Soviet intelligence — the KGB and its military counterpart, the GRU — were even more so.

Today such organizations are just as obsolete in intelligence as they are in the economy. Exactly like General Motors or IBM, the world's major intelligence manufacturers are going through an identity crisis, desperately trying to figure out what went wrong and what business they are really in. And like the corporate dinosaurs, they are being forced to question their basic missions and markets.

Fortunately, like management theorists in the fast-changing business world, a new breed of radical critics is springing up determined not to destroy intelligence but to recast the concept in Third Wave terms.

The very notion of "national security," which these institutions claimed to serve, is being broadened to include not simply military but economic, diplomatic, and even ecological components. A former member of the U.S. National Security Council staff, John L. Peterson, argues that to head off trouble before it explodes the United States should use its intelligence and its military forces to help the world deal with problems like hunger, disaster, and pollution that can throw desperate populations into violent conflict. To do this would require more, not less, intelligence, but different types as well. Again, the parallels with business are striking. Thus, says Peterson, "As the security market moves and broadens, new 'products' will be required to cover the new segments."

Sounding exactly like a business marketing specialist, Andrew

Shepard, a leading CIA analyst and manager, urges intelligence experts to de-massify their output: "To tailor routine intelligence to particular consumers' interests, we need the ability to produce different presentations for each key customer. We envision final assembly and delivery of routine finished intelligence at the 'point of sale.' "

Similarly mirroring Third Wave management thinking, other avant-garde intelligence thinkers speak about listening to "customers," cutting out "middle management," decentralizing, reducing cost, and de-bureaucratizing.

Angelo Codevilla of the Hoover Institution, in Berkeley, suggests that "each part of the government should gather and analyze the secrets it needs." The role of the CIA, he says, should be reduced to that of a clearinghouse. Codevilla urges the United States to retire thousands of spies and spooks stationed in embassies and pretending to be diplomats but collecting information readily available to any informed businessman, journalist, or foreign service officer. The 10 percent of spies operating under diplomatic cover who are useful, he says, should be reassigned to specific government departments, like Defense and Treasury.

More use should be made of part-time informants active in business and professional circles in target countries. If covert operations — foreign operations whose sponsorship can be denied — are needed, they should be carried out by the military or other agencies, not as a part of intelligence.

What's more, Codevilla claims, the technical means of intelligence collection, including some satellite systems, function as indiscriminate "electronic vacuum cleaners," picking up too much chaff along with any wheat. They, like military weaponry, need to be precision-targeted.

The "wheat" that users want is changing, too, even in the military. Thus an influential document circulated at the top of the Pentagon in January 1993 charged that senior military intelligence analysts were "still essentially chewing on" notions of large ground wars. They were focusing too narrowly on military factors, underestimating the importance of political strategy. "Analysts," it declared, "seem to have little feel for or data about the kinds of Third World opposition force we might encounter" and how "militarily insignificant opponents (such as the Serb forces in Bosnia) might pose extremely stressful problems."

NEW MARKETS

According to Bruce D. Berkowitz, a former CIA analyst and Allan E. Goodman, formerly that agency's Presidential Briefing Coordinator, "Rather than detecting and analyzing a jet aircraft which emits a familiar visual, infrared, and telemetry signal . . . the intelligence community may have to detect and analyze old, small aircraft transporting drugs." Rather than spotting tank battalions in movement, it may have to spot guerrillas. And rather than dissecting a Soviet arms-control proposal, it may have to assess a country's attitude toward terrorism.

Fighting terrorism, in particular, requires extremely fine-grained information and new, computerized techniques for getting it. The words of Count de Marenches, former chief of French intelligence, ring true: "Precision personal intelligence can be more critical than precision-guided munitions."

At a March 1993 meeting of AIPASG (the intelligence community's Advanced Information Processing and Analysis Steering Group), Christopher Westphal and Robert Beckman of Alta Analytics described new software to help authorities zero in on terrorist groups by searching out concealed relationships in multiple data bases. Using it, an anti-terrorist squad could, for example, ask the computer to show all locations frequented by six or more selected people. The idea is to let the user "quickly discover and expose critical associations that would otherwise go undetected."

The reasoning is clear. "When vehicles, telephones, or locations are featured in a group, the question must be asked, 'Why is this node here?' and 'Who is the person behind/associated with this node?' " It is claimed the program, called NETMAP, can even locate "emerging" groupings.

Presumably by combining such data with information drawn from bank accounts, credit cards, subscription lists, and other sources, such software can help pinpoint groups — or individuals — who fit a terrorist profile. (Not mentioned in the presentation was the less benign possibility that the same program might help governments pinpoint other, nonviolent political dissidents, mildly oddball religions, or legitimate groups fighting for civil rights.)

At the same conference, Marc R. Halley and Dennis Murphy of the Analytic Sciences Corporation (TASC) proposed software to help track arms sales in the world. The system, they suggested, would collect data about buyers, sellers, items, dates, and quantities. In an era of rising intangibility in warfare, however, it may be equally important to monitor "knowledge factors" like the enemy troops' religious views, culture, time perspective, level of education and training, their sources of information, the media they watch when off duty, and other elements related to knowledge power. In short, knowing the knowledge terrain will be as important for Third Wave armies as knowing the geography and topology of the battlefield was in the past.

THE HUMAN FACTOR

The need for a vast, highly automated network of satellites and sensors to monitor Soviet nuclear and missile development resulted in a de-emphasis on "humint" — the collection of information from human sources. What that meant was a heavy focus on the adversary's capabilities, as distinct from its intentions.

It is true that sometimes the development or deployment of "capabilities" — read tanks, missiles, planes, divisions, and other material elements — can suggest the other side's intentions. But the best satellites can't peer into a terrorist's mind. Nor can they necessarily reveal the intentions of a Saddam Hussein. Satellites and other technical surveillance technologies told the United States that Saddam was massing troops near the Kuwait border. But the United States — short on spies in Baghdad's inner circles — brushed aside such warnings as alarmist and mistakenly concluded the troop movements were just a bluff. One human spy in or near Saddam's inner circle might have cast light on his intentions and changed history.

The shift to a Third Wave intelligence system, paradoxically, means a stronger emphasis on human spies — the only kind available in the First Wave world. Only now, First Wave spies come armed with sophisticated Third Wave technologies.

THE QUALITY CRISIS

The Second Wave stress on mass collection of data by technological means has also contributed to "analysis paralysis." So much chaff has come streaming in from the existing sensors, satellites, and sonars, that it is hard to find the "wheat" mixed with it. Extremely sophisticated software helps scan telephone conversations for keywords. It monitors types and levels of electronic activity, scans for missile plumes, photographs nuclear facilities, and does much else besides. But the analysts have been unable to keep up with the "take" and convert it into timely, useful intelligence.

The result has been an emphasis on quantity rather than quality — exactly the problem faced by General Motors and many other corporations now trying to survive global competition. Because of over-compartmentalization of information, even high-quality analytic "product" frequently failed to reach the right person at the right time. The old system did not provide "just-in-time" intelligence delivery to those who needed it most.

For all these reasons intelligence product has been losing value in the eyes of many of its "customers." Not surprisingly, many users, from the U.S. president on down, simply ignore the classified memos piling up in their in-boxes and the secret briefings they receive. Indeed, secrecy, itself — including the assumptions behind it — is coming under review.

Says a high officer in the Office of the Secretary of Defense, "There was an enormous cult of secrecy — and secrecy itself became a litmus test as to the validity of ideas." If it wasn't secret information, it wasn't important or correct.

In 1992 the U.S. government produced 6,300,000 "classified" documents. The least restricted — not technically classified — bear the stamp "For Official Use Only," otherwise known as FOUO. The next category, which is more restricted and *is* classified, is termed "Confidential." Above that come documents that are "Secret" — some of which are "NATO Secret," meaning they can be shared with other nations who belong to NATO. Others cannot be shared. Then comes "Top Secret" and "NATO Top Secret." But we are only half-way up the mountain so far and still well below the celestial reaches of secrecy. Above "Top Secret" there is a category known as "SCI," or "Sensitive Compartmented Intelligence," open to still fewer people. It is not until we clamber up this peak that we reach information that can only be distributed to so-called BIGOT lists — persons armed with specific code words.

Lest this system seem too simple, it is further matrixed with quali-fiers like "NOFORN," meaning no distribution to foreigners; or "NOCONTRACT," which, not surprisingly, means not to be handed out to contractors; or "WNINTEL," which stands for "Warning Notice — Intelligence Sources or Methods Involved"; or "ORCON," which means "Originator Controls Further Dissem-ination."

This entire dizzying, high-cost edifice is now under sustained attack. When does secrecy increase military strength and when does it, in fact, weaken security? In the words of G. A. Keyworth II, former science adviser to President Reagan, "The price of protecting information is so high that classification becomes a handicap." The new skepticism about secrecy is a direct result of today's Third Wave changes and the competition they have produced.

THE RIVAL STORE

What the Third Wave has done is explosively expand the amount of information (including misinformation) moving around the world. The computer revolution, the multiplication of satellites, the spread of copying machines, VCRs, electronic networks, data bases, faxes, cable television, direct broadcast satellite, and dozens and scores of other information handling and distributing technologies have created many rivers of data, information, and knowledge that now pour into a vast, constantly growing ocean of images, symbols, statistics, words, and sounds. The Third Wave, to switch metaphors, has touched off a kind of informational "big bang" — creating an infinitely expanding universe of knowledge.

This has essentially opened a rival store next door to the spy shop — a Third Wave competitor that makes information available faster and cheaper than the Second Wave intelligence factories. Of course, it cannot supply everything needed by a government or its military. But it can provide a vast amount.

In turn, the Third Wave explosion of information and communication means that more and more of what decision makers need to know can be found in "open" sources. Even a great deal of military intelligence can come from the wide-open store next door. To ignore all this and base analyses on closed sources alone is not only expensive but stupid.

Few have thought as deeply or imaginatively about such questions as a super-smart, forty-one-year-old former Marine and intelligence expert named Robert D. Steele. In 1976 at Lehigh University, Steele wrote his master's thesis on "predicting revolution." Soon he had a chance to find out firsthand what revolution was all about. A tall, chunky man with a booming voice, Steele purportedly served as a political officer in the U.S. embassy in El Salvador during the civil war, although his later career suggests he had intelligence duties in that country. He later returned to Washington, shifted career paths, and became a team leader responsible for the application of information technology to foreign policy issues.

Along the way he graduated from the Naval War College and the Harvard Executive Program in Public Management (Intelligence Policy), and came to represent the Marine Corps on the Foreign Intelligence Priorities Committee and other defense intelligence bodies. Most recently he served as a senior civilian in Marine Corps intelligence, immersing himself in computers, artificial intelligence, and the broader questions of knowledge policy.

Steele wouldn't agree with the *Times* editorialist's throwaway notion that the world is "pretty well covered" by U.S. intelligence. He argues that the United States is, in fact, pitifully short of good linguists, area specialists with actual on-the-ground experience in the areas of their expertise, and even shorter on "indigenous" agents — spies — in critical regions of the world. Nor do Americans, he says, have the patience needed to develop such resources.

Sounding like the new breed of CEO in American business, he complains of organizational short-term-itis. U.S. intelligence, he says, usually places too much emphasis on immediate payback, not enough on long-term nurturing of its secret foreign assets.

Steele takes seriously the new threats posed by today's world. He believes the United States is hopelessly ill-equipped for a reality in which ideological, religious, or cultural warriors roam the planet, and computer "crackers" can turn up in countries like Colombia or Iran, placing their talents at the service of criminals or fanatics.

So Steele doesn't want to shut down U.S. intelligence. Nor does he want the bloated dinosaur shrunk down into a mini-dinosaur. What he calls for, instead, is a profound restructure so that what comes out may be small, or smaller, but will not look like a dinosaur at all.

He believes that much of the U.S. intelligence community will, in fact, eventually disappear down the black hole of budget cuts. A second part, he says, will be privatized. For example, the U.S. Foreign Broadcast Information Service listens to hundreds of foreign radio and television broadcasts and transcribes them for political, diplomatic, and military analysts. Functions like these, he argues, ought to be contracted out to private enterprise. You don't necessarily need government spies to listen to the radio or TV.

A third part of existing intelligence operations — analysis — will be decentralized. Instead of giant pools of analysts working in a central agency, many will be reassigned to work inside government departments like Commerce, Treasury, State, or Agriculture, as has been suggested by Shepard, Codevilla, and others, tailoring analysis on the spot to the needs of the users.

But none of this is central to Steele's one-man campaign. He has, as it were, a bigger whale to harpoon — the Leviathan of secrecy. Indeed, Steele may well be the single most forceful enemy of secrecy in Washington.

"If there is a terrorist group and it has a biotoxin that could cause a catastrophe and you have managed to plant an agent in the group, of course, you need to keep his identity secret. Of course, some secrets are necessary. But the hidden costs of secrecy are so immense they often outweigh the benefits by a wide margin," Steele contends.

For example, armies like to keep their "deficiencies" secret so that the enemy can't target their weaknesses. But the same restrictions that keep the enemy ignorant often deny information to the very people who might fix the deficiency. So weaknesses are discovered late if at all. Because information is compartmentalized in the interests of secrecy, different groups in an agency pursue different solutions to similar problems, and the information they develop is harder to synthesize, disseminate, and utilize. Worse yet, Steele argues, the analysts are cut off from the external world and live in what he calls "virtual unreality."

One of the things the Marine Corps did while Steele was a senior civilian in its intelligence arm was to give SPARC workstations to its analysts. The computers provided them instantaneously with the highest-level secret material. But the Marines also built a separate small glass-walled room nearby and put an ordinary PC into it. Using that machine, an analyst could link up with Internet to access thousands of data bases around the world — all filled with open, publicly available, nonsecret information. The analysts discovered to their surprise that much of what they needed to know could not be found in the secret material. Because of secrecy requirements, their workstations were not hooked up to open or public networks. As a result they turned to the modest little PC, which was connected to the world outside, and they found much of what they needed in easily available open material.

Steele became so convinced of the intelligence value of open source information that he talked the Marines into allowing him, on his own time, and at his own expense, to organize what became the first Open Sources Symposium — a conference held in Virginia in November 1992. The ironic play on the initials of the Office of Strategic Services (forerunner of the CIA) could not have been lost on his audience and speakers who included the chief of staff of the Defense Intelligence Agency, a former science adviser to the president, the deputy director of Central Intelligence, and a surprising mix of people from the information industry, as well as members or observers of the far edge of the computer hacker community. Present also were John Perry Barlow, lyricist for the Grateful Dead, and Howard Rheingold, author of *Virtual Reality* and *The Virtual Community*.

It is unlikely that anyone less committed to the concept of open sources, less brash, or less bound by military and intelligence community convention could have pulled off such an event. But Steele is driven by a vision that reaches far beyond the immediate.

"Imagine," he exhorted that first Open Sources Symposium, "an extended network of citizen analysts, competitive intelligence analysts in the private sector, and government intelligence analysts — each able to access the other, share unclassified files, rapidly establish [computer] bulletin boards on topics of mutual interest, and quickly pull together opinions, insights, and multimedia data which is all the more valuable for being immediately disseminable without restriction. This is where I think we need to go." He wants intelligence to draw on all the "distributed" knowledge available in society.

But even this does not capture the breadth of his vision. Steele wants more. He proposes to "link national intelligence with national competitiveness . . . , making intelligence the apex of the knowledge infrastructure." He not only believes intelligence should draw on public sources but that it should also, for the most part, be made available to the public. He speaks of using intelligence to provide valuable information "from schoolhouse to White House."

Steele sees "intelligence as part of a continuum, or a larger national construct, which must also include our formal educational process, our informal cultural values, our structured information-technology architecture, our informal social and professional networks for information exchange, our political governance system." He sees intelligence, in short, not just as a source of cloak-and-dagger information massaged into "estimates" for a handful of top policymakers but as a vibrant contributor to the knowledge system of society as a whole.

Steele's vision will thrill many — and send a nervous quiver down the spines of others. It has cracks and unfilled gaps in it that critics may be quick to seize on. His direct manner may put people off. And his dream, like most dreams, is unlikely to be fully realized. But it positions intelligence within a vastly larger framework than any previously discussed. His campaign is one of the forces aimed at adapting intelligence to the realities of the Third Wave.

To worry about war or anti-war in the future without rethinking intelligence and seeing how it fits into the concept of knowledge strategy is an exercise in futility. The restructure and reconceptualization of intelligence — and military intelligence as a part of it — is a step toward the formulation of knowledge strategies needed either to fight or forestall the wars of tomorrow.

18

SPIN

T HE PEOPLE thinking hardest about warfare in the future know that some of the most important combat of tomorrow will take place on the media battlefield.

Just as the United States cannot develop a fully comprehensive knowledge strategy until it puts its intelligence house in order, it faces an even greater problem with respect to the media. Thus, according to Neil Munro of *Defense News*, the U.S. military will run into a "brick wall" because the Department of Defense has only limited authority to involve itself in the media. The American Constitution, as well as its culture and politics, sets limits on censorship, and "propaganda" is a dirty word to most Americans.

Thus, while the military knows that putting the right "spin" on war news can, at times, be as important as devastating an enemy's tanks, nobody loves a "spin doctor" who wears khaki. Especially the American press.

After the Gulf War a fiery dispute broke out between the American media and the Pentagon over its attempts to manage the news and its deliberate effort to keep reporters away from ground combat. But, as intense as that confrontation may have seemed, temperatures are likely to rise still further in the years to come. Knowledge strategists will have to take this into account.

THE GERMAN MEDAL

Propaganda, writes historian Philip Taylor, "came of age under the ancient Greeks." But it came of age again after the industrial revolution gave rise to the mass media. Thus the Second Wave war-form was accompanied by one-sided news, doctored photographs, and what the Russians call "maskirovka" (deception) and "dezinformatsia" (disinformation) transmitted through the mass media. Tomorrow, as the Third Wave war-form develops, propaganda and the media that convey it will both be revolutionized.

To know how "spin" is applied we need to recognize the different levels at which the military propaganda game is played. At the strategic level, for instance, adroit propaganda can actually help make or break alliances.

During World War I both Germany and Britain each tried to draw American support. The British knowledge warriors were far more sophisticated than the Germans and seized on every symbolic event to paint the Germans as anti-American. When a German U-boat torpedoed the American ship *Lusitania*, which we now know may have been carrying munitions to the British, American opinion was outraged. But the real outrage was orchestrated a year later by the British.

On discovering that a German artist had made a bronze medal to celebrate the sinking of the ship, the British stamped out replicas of the medal, boxed them, and sent hundreds of thousands of them to Americans along with an anti-German propaganda leaflet. In the end, of course, America did enter the war on the British side, dooming the Germans. The decision, based on American financial and other interests at the time, cannot be attributed to British propaganda alone. But strategic propaganda helped make the decision palatable to the American public.

More recently, in the Gulf War, President Bush's effective mobilization of United Nations support was accompanied by propaganda suggesting that the United States, rather than acting in its own interest, was merely doing the UN's bidding. The strategic purpose of this campaign was to isolate Iraq diplomatically, and it succeeded.

Propaganda can also be conducted at the operational or theater level. Saddam Hussein's regime was aggressively secular, not Islamic, but his information ministry continually played the Islamic card, picturing Iraq as the defender of the faith and the U.S.-backed Saudi Arabia as a traitor to the religion.

Finally, at the tactical level, U.S. psychological warfare specialists dropped 29 million propaganda leaflets with thirty-three different messages over Iraqi troops in Kuwait, providing instructions on how to surrender, promising humane treatment of prisoners, encouraging them to desert their equipment, and warning them of attacks to come.

Smart spin doctors know exactly whether their goals are strategic, operational, or tactical and work accordingly.

SIX WRENCHES THAT TWIST THE MIND

Khaki-clad spin doctors have used six tools over and over again through the years. These are like wrenches designed to twist the mind.

One of the most common is the atrocity accusation. When a fifteen-year-old Kuwaiti girl testified before Congress during the Gulf War to the effect that Iraqi troops in Kuwait were killing premature babies and stealing the incubators to take them back to Iraq, she twanged many a heartstring. The world was not told that she just happened to be the daughter of the Kuwaiti ambassador in Washington and a member of the royal family, or that her appearance was stage-managed by the Hill & Knowlton public relations firm on behalf of the Kuwaitis.

Of course, propaganda need not be false. Widespread accounts of Iraqi brutality in Kuwait were confirmed when reporters arrived after the Iraqis were driven out. But atrocity stories, both true and false, have been a staple of war propaganda. In World War I, writes Taylor in his excellent history of war propaganda, *Munitions of the Mind*, Allied propagandists constantly invoked "Images of the bloated Prussian 'Ogre' . . . busily crucifying soldiers, violating women, mutilating babies, desecrating and looting churches."

Half a century later, atrocity stories were important in the Vietnam War, during which accounts of the My Lai massacre by American soldiers disgusted wide sectors of the American public and fed the anti-war fervor. Atrocity stories, both true and false, filled the air during the Serb-Bosnian conflict.

A second common tool is hyperbolic inflation of the stakes involved in a battle or war. Soldiers and civilians are told that everything they hold dear is at risk. President Bush pictured the Gulf conflict as a war for a new and better world order. At stake was not simply the independence of Kuwait, the protection of the world's oil supply, or elimination of a potential nuclear threat from Saddam, but, supposedly, the fate of civilization itself. As for Saddam, the war was not about his failure to pay back billions of dollars borrowed from the Kuwaitis during the earlier Iran-Iraq war; it was — he claimed — about the entire future of the "Arab Nation."

A third mind-wrench in the military spin doctor's kit bag is demonization and/or dehumanization of the opponent. For Saddam as for his enemies in next-door Iran, America was "the Great Satan," Bush was "the Devil in the White House." In turn, for Bush, Saddam was a "Hitler." Baghdad radio spoke of American pilots as "rats" and "predatory beasts." An American colonel described an air strike as "almost like you flipped on the light in the kitchen at night and the cockroaches start scurrying there, and we're killing them."

A fourth tool is polarization. "Those who are not with us are against us."

A fifth is the claim of divine sanction. If Saddam draped his aggression in Islamic garb, President Bush also called upon God's support. As the Moroccan sociologist Fatima Mernissi has pointed out, the incantatory phrase "God bless America" ran through American propaganda — and had an odd, unanticipated side effect when it reached ears in the souks and streets of the Muslim world. Accustomed to regarding America as the apostle of materialism and atheism, ordinary people in the streets of North Africa and the Middle East were, she says, "agog" when Bush invoked God. Did Americans actually believe in God? The confusion was even greater when God was linked with rhetoric about democracy. Was democracy a religion?

Finally, perhaps the most powerful mind-wrench of all is meta-propaganda — propaganda that discredits the other side's propaganda. Coalition spokespeople in the Gulf repeatedly and accurately pointed out that Saddam Hussein had total control of the Iraqi press and that, therefore, the people of Iraq were denied the truth and Iraqi airwaves were filled with lies. Meta-propaganda is particularly potent because, instead of challenging the veracity of a single story, it calls into question everything coming from the enemy. Its aim is to produce wholesale, as distinct from retail, disbelief.

What is striking about this entire list of military propaganda techniques is its Second Wave character. Each of these "mind-wrenches" is designed to exploit the mass media to sway mass emotion in mass societies.

NEO-NAZIS AND SPECIAL EFFECTS

These "classical" instruments of the spin doctor may continue to work in conflicts between countries with centralized Second Wave media. The same tool can be exploited by Third Wave societies against Second Wave societies. But in Third Wave societies, the media revolution is rewriting all the rules.

To begin with, Third Wave economies develop a vast multiplicity of channels through which both information and misinformation may pour. Cellular telephones, PCs, copying machines, fax, videocams, and digital networks permit the exchange of vast volumes of voice, data, and graphic material through multiple, redundant, and decentralized channels, often out of easy reach of government or military censors.

Thousands of computer-based "bulletin boards" are also springing up, linking millions of individuals around the world in a continuing conversation about everything from sex to stock market tips to politics. Such systems are mushrooming at high speed, crossing national boundaries, and facilitating the formation of groups devoted to everything from astrology, music, and ecology, to neo-Nazi paramilitary operations and terrorism. The overlapping and interlinked networks on which these systems depend are almost impossible to stamp out. Given the proliferating new media, crude centralized propaganda pumped down from above may increasingly be countered from below.

These new media tend to disperse power. A single videotape, shot by an amateur, of Los Angeles police brutalizing a black man, led to riots that caused almost as many casualties and as much damage as a small war. Videocameras are increasingly used to document abuses of power by local and national governments. And they are circulated, if not on TV, then in the form of videocassettes. Central control is weakened by the new media. It will be further debilitated by interactivity that will permit users to talk back to the central authorities. Radio talk shows and home shopping via TV are pale foreshadowings of this process.

The TV set will eventually be replaced by a (possibly wireless) unit that will combine a computer, a scanner, a fax, a telephone, and a desktop tool for creating multimedia messages all rolled into one and networked to one another. And instead of keyboards, these "telecomputers" may eventually be operated by speech commands in natural language.

What all this points toward is a world in which millions of individuals have, at their command, the power to create Hollywood-like special effects, virtual-reality-based simulations, and other potent messages — power that not even governments and movie studios had available in the past. The world will be divided, as it were, into preelectronic communities so poor that even television sets are scarce; communities in which conventional broadcast television is essentially universal; and networked communities in which conventional broadcast television, as we know it, has been left behind.

THE MEDIA AS "STAR"

When we look back at the Gulf War, the first in which elements of Third Wave warfare were used decisively, we find that, in a sense, the war may not have been the point of all the media coverage. The media, itself, became the "star" of the spectacle. As former Maj. Gen. Perry Smith, himself a CNN personality, noted, "Over the six weeks of the war more people watched more hours of television per day than at any time in history."

Impressive as that may seem, other changes are even more important. The media are fusing into an interactive, self-referencing system in which ideas, information, and images flow incestuously from one medium to another. TV clips of war news, for example, suggest stories to newspaper editors; movies about the military, like *A Few Good Men*, generate printed commentary, radio, and TV interviews; TV sitcoms picture journalists at work; newsphotos shot (or staged) on the battlefield for a newsmagazine turn up as a TV clip. All increasingly rely on computers, fax machines, satellites, and telecom networks and merge to form an integrated or fused media system.

In this embryonic system, television (for now, but only for now) sets the news agenda, especially in war coverage. While some American TV news directors may still check headlines in the *New York Times* or the *Washington Post* before deciding which political or diplomatic stories to feature, in most other matters the influence of print is declining.

"With the Gulf War," writes Ignacio Ramonet in *Le Monde Diplomatique*, television "has seized power," shaping the style, and above all the rhythm and pace of print journalism. TV has succeeded in imposing itself on the other media, Ramonet points out, "not only because it presents a spectacle, but because it has become faster than the others." We will return to this crucial insight in a moment. Before that, however, we need to ask how military propagandists might adapt to the arrival of Third Wave communications.

THE PINPOINT MESSAGE

Some things are clear. Precision-targeting information is just as important as precision-targeting weapons, and the new media will make this possible to an unprecedented degree.

When targeting audiences in Third Wave societies, tomorrow's media manipulator, like the ad agencies of tomorrow, will have to de-massify the messages, crafting different versions for each audience segment — one for African-Americans, another for Asians, still another for doctors, and another for single mothers, as the cases may be. Fake atrocity stories will, no doubt, someday be engineered in this way, with "victims" described differently in each version, so as to generate maximum sympathy or hatred for each set of viewers.

Such segmentation, however, is only a half step toward the ultimate goal: individualization. Here each message will be massaged to produce maximal impact on one person, rather than a group. The "Dear Mary" approach of the direct-mail copywriter today will be developed and extended, using multiple commercial and government data bases to extract a profile of the individual. Armed with data from credit card tax records and medical secrets a spin doctor could eventually surround a targeted individual with coordinated, personalized subtle messages via print, television, videogames, data bases, and other media.

Propaganda for and against war, often originating from senders halfway across the world, sometimes masking the real source, will be cleverly infiltrated into the news exactly as entertainment is infiltrated into it today. Ordinary entertainment programs, too, may be altered to contain concealed propaganda messages tailored for each individual or family.

Seemingly impossible and costly today, this ultimate customization of communication will become quite feasible when Third Wave media and telecommunications systems are fully developed.

REPORTING IN REAL TIME

This shift toward total de-massification will be accompanied by a further acceleration into real time. And this will intensify conflict between the military and the media.

In 1815 two thousand American and British soldiers killed one another in the Battle of New Orleans because news of a peace treaty signed two weeks earlier in Brussels didn't reach them in time. News moved at a glacial pace.

With industrialization, it accelerated, but it still moved at pre-electronic speeds. An outgrowth of the rise of the mass media was a new profession — "war correspondent." Many combat journalists — Winston Churchill, who rode with the British troops in the Boer War and later became Britain's great wartime prime minister . . . Richard Harding Davis in the Spanish American War . . . Ernest Hemingway, who chronicled the life of the Loyalists in the Spanish Civil War . . . Ernie Pyle in World War II — subsequently became legends in their own time. But by the time their dispatches were printed, the battles they described were already over. Their reports from the field could not influence the actual battlefield outcomes.

Today battles and peace treaties become news before they are concluded. By the time U.S. forces arrived in Somalia, an army of TV cameras were on the beach to greet them. Presidents and prime ministers learn what is happening from TV before diplomats can report back to them. Leaders send messages to one another not simply through ambassadors, but directly on CNN, confident that their counterparts and adversaries will be watching — and will, in turn, respond on camera.

During Iraq's Scud missile attacks on Tel Aviv, Israeli military censors knew that CNN was being closely monitored in Baghdad. They worried that CNN pictures showing where the missiles were hitting would help the Iraqis zero in on targets more accurately. The sheer acceleration of news had changed its significance.

Writing on "Information, Truth and War," Col. Alan Campen notes that "satellite technology makes moot the issue of censorship." Commercial reconnaissance satellites will make it almost impossible for combatants to hide from the media, and with all sides watching the video screen, instant broadcasts from the battle zone threaten to alter the actual dynamics and strategies in war. It can, Campen says, "transform reporters from dispassionate observers to unwitting, even unwilling, but nonetheless direct participants" in a war.

Campen argues that the citizens in a democracy may have both a right and a need to know what is going on. But, he asks, do they need to know it in real time?

UNREAL REAL TIME

The new media change not merely reality, but even more important, our perception of it — and, therefore, the context in which both war and peace propaganda contend. Before the industrial revolution, peasant populations were illiterate and provincial, relying on travelers' tales, church dogma, or myth and legend for their images of events distant in time or place. The Second Wave mass media brought distant places and times into closer focus and gave a "you are there" quality to what purported to be news. The world was pictured as objective and "real."

By contrast, Third Wave media are beginning to create a sense of unreality about real events. Early critics of television lamented its immersion of the viewer in a vicarious world of soap opera, canned laughter, and false emotions. These concerns will seem trivial tomorrow, for the new media system is creating an entirely "fictive" world to which governments, armies, and whole populations respond as though it were real. In turn, their actions are then media-processed and plugged into the fictional electronic mosaic that guides our behavior.

This growing fictionalization of reality is found not only where it belongs, in sitcoms and dramas, but in news programming as well, where it may promote the deadliest of consequences. This danger is already being discussed around the world.

The Moroccan newspaper *Le Matin*, in Casablanca, recently carried a thoughtful essay quoting the French philosopher Baudrillard, to the effect that the Gulf War came across as a gigantic simulation, rather than a real event. "Media-tization," the newspaper agreed, "reinforces the fictive character" of events, making them seem somehow unreal.

VIDEO ON VIDEO

This irreal quality was amplified during the Gulf War by what amounted to television of television — TV^2, as it were. Again and again, one saw video images of video screens showing targets and hits. So important was media imagery considered by the military that, according to one U.S. Navy commander, pilots in actual combat sometimes reset their cockpit video displays to make them show up better on CNN. Some weapons, too, it turns out, were more telegenic than others. Thus HARM missiles zero in on enemy air defenses and fire tiny pellets at them. But the damage they do doesn't show up so well on television. What the cameras want are big bomb craters on the runways.

New technologies for simulation make it possible to stage fake propaganda events with which individuals interact, events that are intensely vivid and "real." The new media will make it possible to depict entire battles that never took place or a summit meeting showing (falsely) the other country's leader rejecting peaceful negotiation. In the past, aggressive governments sometimes staged provocations to justify military action; in the future they may only have to simulate them. In the fast-onrushing future, not merely truth but reality itself may be a casualty of war.

The brighter side to all of this is that a public accustomed to using simulation for many other purposes, in the home, at work, and at play, may learn that "seeing" or even "feeling" is not believing. The public is likely to grow increasingly media-sophisticated as time goes by — and hopefully more skeptical as well.

Finally, it is necessary to disabuse ourselves of the by-now conventional notion that the new media are going to homogenize the world, eliminating differences and giving immense, unchallenged influence to a few — that CNN, for example, is going to jam Western values and American propaganda down 5 billion throats.

CNN's current dominance in the worldwide TV news market is temporary, for rival networks are already in formation. Within a decade or two we can expect a multiplication of global channels, paralleling the diversification of media already taking place inside the Third Wave countries.

Tiny satellite dishes in homes around the world will someday pick up the evening news from anywhere and everywhere — Nigeria or the Netherlands, Fiji or Finland. Automatic translation will eventually mean that a German family might watch a game show from Turkey automatically translated into German. Orthodox Catholics in the Ukraine may be bombarded by messages from a Vatican satellite calling on them to leave their church and become Roman Catholics instead. Ayatollahs in Qum may be preaching to homes from Kyrgyzstan to the Congo, or, for that matter, California.

Instead of a handful of centrally controlled channels watched by all, vast numbers of humans will eventually gain access to a dazzling variety of over-the-border messages their political and military masters may not wish them to hear or see. Before long, one may assume, the spin doctors and knowledge warriors of many nations, not to mention terrorists and religious fanatics, will begin thinking creatively about how to exploit the new media.

Policies dealing with the regulation, control, or manipulation of the media — or for the defense of freedom of expression — will form a key component of the knowledge strategies of tomorrow. And their knowledge strategies, in turn, will determine how different nations, nonnational groups, and their armies fare in the looming conflicts of the twenty-first century.

In defining or implementing a knowledge strategy, the U.S. military does not have a free hand. First Amendment guarantees of press freedom mean that U.S. spin doctors have to be more subtle and sophisticated than those of countries in which totalitarian control of the media is still a fact.

Yet, despite the Pentagon's frustrations and tensions vis-à-vis the media, and vice versa, most of the military knowledge-warriors to whom we spoke agreed with the media on one essential. They believe that totalitarian control of the media is itself a losing strategy and that, in general, America's tradition of relatively open information pays off militarily.

Many, in and out of uniform, argued earnestly that whatever advantages a totalitarian state might gain by its control of the media are decisively outweighed by the innovativeness, initiative, and imagination that springs from an open society. Having a knowledge strategy, they would say, does not imply imposing totalistic control. It means using the inherent advantages of freedom to better purpose.

But win, lose, or draw, the media, including channels and technologies unimagined today, will be a prime weapon for Third Wave combatants in both the wars and anti-wars of the future, a key component of knowledge strategy.

So far in these pages we have traced the birth of a new war-form that reflects the new way of creating wealth. We saw its origin in the first formulations of AirLand Battle doctrine. We saw that doctrine applied in limited and modified form in the Gulf War. We examined new technologies, like robotics and non-lethal weaponry, likely to be incorporated into the new war-form. And, finally, we have peered ahead at the "knowledge strategies" that the military leaders of tomorrow will need to avert defeat or attain victory in the wars of tomorrow. We have tracked, in other words, an historical progression leading toward the dominant war-form of the early twenty-first century.

What we have not yet explored are the dangers that confront us as a result of the emergence of the Third Wave war-form.

PART FIVE

DANGER

19

PLOUGHSHARES INTO SWORDS

ONE OF THE THINGS the introduction of a new war-form does is profoundly upset existing military balances. That is exactly what happened in the past when, on August 23, 1793, an embattled France, bloodied by revolution and about to be torn apart by invading troops, suddenly imposed universal conscription. The words of the decree were dramatic:

"From this moment . . . all Frenchmen are in permanent requisition for the service of the armies. The young shall fight; the married men shall forge arms and transport provisions; the women shall make tents and clothing and serve in hospitals; the children shall turn old linen into bandages; the aged shall betake themselves to public places to arouse the courage of the soldiers. . . ."

This levee introduced mass warfare into modern history, and was soon combined with innovations in artillery, tactics, communication, and organization, thus giving rise to a powerful new way of making war. Within twenty years France's draftee army, now led by Napoleon, had overrun Europe and marched all the way to Moscow. On September 14, 1812, Napoleon could actually see that city's golden domes sparkling in the sunlight.

Napoleon still had to contend with British sea power. But on the continent his was the only military force that mattered. Europe had gone from a "multi-polar" to a "uni-polar" structure of power.

The Second Wave war-form, then still in embryonic form, could not guarantee victory when, as in the case of the Russian campaign,

Napoleon's supply lines were overextended. Nor could it be used to subdue guerrillas in Spain. But its efficacy was so clear that first Prussia, then other European armies, moved to adopt and further develop many of the French innovations.

Historical analogies are always suspect. Nevertheless, certain similarities between Napoleon's world and our own should give us pause. The United States, too, in introducing a new war-form into history, has radically upset the existing balance of military power, this time not on a single continent but around the globe. Its increasingly Third Wave military tilted the balance so decisively that the Soviet forces in Europe lost their parity with those of the United States and NATO. The combination of the West's knowledge-intensive military, backed by fast-growing knowledge-intensive economies, made the difference that led, ultimately, to the collapse of communism. America emerged as the sole superpower on earth. And the result was once again, a unipolar system.

The actual application of the Third Wave war-form in the Gulf, even in partial and modified form, proved its effectiveness for all to see. And again, like Prussia in the aftermath of the Napoleonic Wars, armies all over the planet today are trying to imitate the United States to the extent possible.

From France, Germany, and Italy to Turkey, Russia, and China the exact same words crop up in their announced plans: rapid deployment ... professionalization ... better electronic air defense ... C³I ... precision ... less reliance on conscription ... combined operations ... interdiction ... smaller forces ... special operations ... power projection ...

Japan, South Korea, Taiwan, and other Asian nations specifically cite the Gulf War as the reason they now prefer better (read information-intensive) technology to bigger forces. The French army's chief of staff, Gen. Amedee Monchal, says that "in 10 years the land forces will lose 17 percent of their troops." By contrast, "The emergence of Electronic War translates into a 70 percent increase" in troops devoted to EW activity. With only a limited grasp of its implications, nations everywhere are preparing, as best they can, to exploit knowledge-intensivity.

Nor are the present perceived limitations of Third Wave war necessarily permanent. After the Gulf conflict conventional wisdom held that the new-style combat would not work in Vietnam-like jungles or Bosnian mountains. "We don't do jungles, and we don't do mountains" became a semi-facetious catchphrase among top-ranking U.S. officers.

As one Pentagon officer, referring to the Balkan conflict, put it in correspondence to us: "Our precision guidance is good, but not enough to hit an individual mortar tube pointed at a village; our ordnance is good, but too big to destroy only the mortar tube without collateral damage to the people and villages we are trying to protect; and we do not have anything like the targeting information necessary to surveil [a] few hundred small and mobile potential targets across the rugged Balkans terrain."

Yet war-forms evolve, technologies improve, and exactly as in the case of post-Napoleonic armies, steps are being taken to overcome the early limitations of the new war-form. As noted in the preceding discussion, the thrust of change is toward strengthening low-intensity combat capabilities with new improved technologies — sensors, space-based communications, non-lethal and robotic weapons. Which suggests that the new, Third Wave war-form may in time prove to be just as powerful against guerrillas and small-scale opponents waging First Wave war as against Iraq-style Second Wave armies.

The emergence of the Third Wave war-form is forcing all governments to reevaluate their military alongside their perceived threats. Today China still has some 3 million men under arms (down from more than 4 million in 1980). Its 4,500 combat aircraft give it the third-largest air force in the world. But China's leaders know that, apart from assuring internal security, their large and costly Second Wave army is no great bargain. And they know that their planes are mostly obsolete — meaning they are not "smart" enough. China looks appraisingly at its neighbors and it is now clear that, in the absence of nuclear weapons, North Korea's million-plus, Soviet-style army is weaker than it looks, while South Korea's 630,000-man American-style army is stronger than it looks. The 246,000-man Japan Self Defense Force, with its great surge capacity and technical skills, is far more powerful than its size alone might suggest.

What should disturb those of us concerned with guarding peace is not raw military power as such, but today's sudden, erratic tilts and changes in relative strengths. For nothing is more likely to increase unpredictability and worst-case paranoia on the part of political leaders and military planners. All of which is heightened by uncertainty about America's military future.

The Napoleon analogy, if nothing else, compels us to consider the transience of power. On June 18, 1815, less than three years from the date of his furthest advance toward the East, Napoleon's empire collapsed at Waterloo. France's "unipolar" moment — its position as a superpower — was over in a brief flick of time. Could the same thing happen to America? Is America's unipolar moment also a flash in the historical pan?

BUDGET WITHOUT STRATEGY

The answer will depend in part on its own actions. To keep a military edge, the United States must also keep its economic edge. It still, despite the rise of Japanese and Asian economies, retains many advantages in science, technology, and other fields. It needs to accelerate the shift out of its residual Second Wave industries while minimizing the social dislocation and unrest that accompany so deep an economic transformation. But it must also rethink its strategic options in fresh ways.

Unfortunately for all concerned, friends and enemies alike, American elites, both political and military, are deeply disoriented not only by the end of the Cold War, but by the split-up of the Western alliance, the economic rise of Asia, and, above all, by the arrival of a knowledge-based economy whose global requirements are by no means clear to them.

The result is a dangerous lack of clarity about America's long-term interests. In the absence of such clarity, even the best armed forces in the world could, in the future, be sent to defeat or — worse yet — to die for trivial or peripheral purposes. With congressional budget-butchers, moreover, chopping away at Pentagon funds with little understanding of the Third Wave war-form, the United States lead could, in fact, quickly dissipate.

In a logical world it is impossible to know how big a military budget a country needs until the country has a strategy and can assess its requirements. But that is not the way military budgets are arrived at. As former U.S. secretary of defense Dick Cheney once told us, in the real world "budget drives strategy, strategy doesn't drive budget."

Worse yet, the budgets that do the "driving" are not determined in a remotely rational fashion, either. In every country, arms and armies are the ultimate political pork barrel, providing jobs, profits, and payoffs. Domestic political power and interservice rivalries, not logic, drive the budget process. Thus current arguments over the size of the defense budget are essentially ammunition for different constituencies in their claims on government money, rather than genuine strategic debates.

But even more dangerous than myopic budget-bashing and strategic bewilderment — and dangerous not just for the United States — is today's ill-perceived transformation in the relations between the economy and the military, between wealth, that is, and war.

MERCHANTS OF DEATH

Throughout the Second Wave era the armed might of major powers was backed by large-scale defense industry. Massive naval shipyards served the world's Second Wave navies. Vast companies arose to produce tanks, planes, submarines, munitions, and missiles.

For generations peace advocates, in turn, relentlessly attacked the arms industry. Excoriated as "merchants of death," or a "subterranean conspiracy against peace," the munitions makers of the world were pictured, sometimes with justification, as fanning if not actually sparking the flames of war.

"Take the profits out of war" became a familiar slogan. Books like *Bloody Traffic*, published in 1933, and its successor version, *Death Pays a Dividend*, published in 1944, exposed corruption and warmongering by what later came to be known as the "military industrial complex."

Today, it might seem, critics of that complex might take heart — the defense industries are in mortal trouble. The number of workers employed by the war industries is plummeting in the high-tech nations (though not in some of the smaller, poor nations). In the United States, daily headlines report the layoffs of scientists, engineers, technicians, and less-skilled defense workers. General Dynamics, for example, maker of fighter planes and submarines, laid off 17,000 workers in twenty months. In the United States as a whole, with many military factories standing vacant, some 300,000 defense jobs vanished in less than two years after the fall of the Berlin Wall, and many more have followed since.

Thrashing about and desperate to survive, the giant defense firms are restructuring, merging, and casting about for new business. But even if they manage to dodge the current volley of budget-bullets, the military industries suffer from a long-term illness. Many firms will perish. Yet the chances for peace may actually be worsened as a consequence. For what the world now faces is the civilianization of war and weapons.

In one of the great ironies of history, those who worked so hard and selflessly to promote a downsizing of the defense industry, hoping to shift military expenditure to presumably more benign purposes, are hastening this civilianization. And that, it now turns out, will spark new, ill-recognized dangers for the world.

THE CIVILIANIZATION OF WAR

By "civilianization" we do not mean conversion or the beating of swords into ploughshares. We mean, rather, its opposite, the transfer of militarily relevant work once carried out by military-specific industries to civilian-oriented industries instead.

A great deal of attention has been lavished on a small number of examples of conversion, like the joint Lockheed-AT&T venture to automate highway tollgates with "smart cards," or the effort at Lawrence Livermore National Laboratory to build computer models of climatic change using work initially devoted to the study of nuclear explosions. Thomson-CSF, the French defense giant, has applied some of its military electronics know-how to building a network for France Telecom, the phone company.

But even as politicians and the media in various nations extol the blessings of conversion, a far more extensive counter-process is converting civilian industries to wartime capabilities. This is civilianization. It is the real "conversion." And what it does is the opposite of what was initially intended: it beats ploughshares into swords.

Civilianization will soon give fearsome military capabilities to some of the smallest, poorest, and worst-governed nations on earth. Not to mention the nastiest of social movements.

TWO-FACED "THINGS"

The main purpose of a military industrial complex in any country was to turn out things called "arms" — products specifically designed to kill or to support killing, from rifles and grenades to nuclear warheads. There were always, of course, some "dual-use" products that were primarily created for civilian purposes, then later used by the military. Trucks that could take barrels of milk from farm to city might take ammunition to the front instead. But with the exception of food and oil, Second Wave wars were not won with consumer products.

What happens, however, if that consumer product is, in fact, a supercomputer capable of designing nuclear weapons? Or how about the cable television box that sits in millions of American homes — and contains highly sophisticated encryption technology potentially useful in missile guidance? Or extra-sensitive detonators and pulse lasers? Or myriad other products made for the civilian economy?

In a Third Wave world, in which both technologies and products diversify to meet the demands of de-massified markets, the number of items with dual-use potential grows. And when we look beyond products and technologies to their components and subtechnologies, the number of potential military permutations skyrockets. For this reason, says one defense analyst, armies of the future will "swim in the sea of civil technology."

In turn, the very diversity of products and technologies translates into a far greater diversity of weapons as well. The rise of knowledge-intensive, high-tech economies is also marked by the multiplication of marketing channels, the liberalization of capital flows, and the rapid movement of people, goods, services, and especially information across increasingly porous borders. All this means that dual-use items flow more easily through the global bloodstream.

But to focus exclusively on specific dual-use "things" is to miss the broader point. Not only goods but services are involved. And not just here on earth, but in space as well.

CONSUMER SERVICES FOR WAR

Listen to defense consultant Daniel Goure, former Director of Competitive Strategies in the Office of the U.S. Secretary of Defense. We face, he says, "a global revolution in terms of access to space communications, surveillance, and navigation, all elements critical to military capability."

Consider surveillance. "A future Saddam Hussein," Goure says, "will be able to subscribe to the information stream from a dozen or so surveillance sensors, of various kinds and qualities, Russian, French, Japanese, possibly even from the United States itself. All commercial."

Even now the Russian Nomad system, once called Almaz, makes surveillance imagery commercially available with a resolution down to about five meters. "For precision targeting," Goure notes, "you'd like one meter. But, frankly, the civil technology [available to any buyer] is better now than our military had in the 1970s, and we thought we were pretty spiffy."

Almost any government, therefore, anywhere in the world —
including the most fanatic, aggressive, repressive, and irresponsible — may soon be able to buy eyes in the skies to provide sophisticated images of U.S. tanks or troops or missile emplacements to
within about fifteen feet of accuracy. Coming refinements in navigation technology will soon give positional information down to little
more than one meter. Even though U.S. satellites can now provide the
highest precision, America's dominance in space could, for all practical purposes, be neutralized.

This is not all. Space also provided the allies with advanced communications during the Gulf War. But Motorola today is planning to
put a ring of satellites around the earth. This commercial system,
called Iridium, could provide essentially unjammable communications to users anywhere. Moreover, as electronic networks proliferate
on the ground, it will soon be impossible to deny a future opponent
access to satellite-based intelligence. Critical battlefield information
could flow down to commercial ground stations and data bases in
Zurich, Hong Kong, or São Paulo and be fed through any number of
intermediary networks to armies in, say, Afghanistan, Iran, North
Korea, or Zaire. Such information can be used, among other things, to
target and guide missiles.

"SMART" VERSUS "SMARTENED" ARMIES

Then there are the missiles themselves. Tomorrow's Saddam Hussein,
Goure notes, will have "the ability to take relatively old technology,
like a Scud missile, and . . . put it down precisely on a target. All you
need to do is add a commercial GPS navigational receiver like the
Slugger, of Gulf War fame, plus some rewiring and some other items,
and for around five thousand dollars in, say five years, Saddam or the
Iranians or anyone else could have a smart Scud" — instead of the
notoriously wobbly and hard-to-target Scud launched against Tel
Aviv and Riyadh.

In short, adding commercially available Third Wave "smarts" to
old, Second Wave weapons can transform them into intelligent
weapons at peanut prices that even impoverished armies can afford.
Thus today's smart armies will find themselves faced by tomorrow's
smartened armies.

It is true that the United States and other militarily advanced nations retain certain advantages — better-educated troops, more
rounded capabilities, and better systems integration. But the lop-

sidedness of the Gulf War is unlikely to be repeated in the future, as some elements, at least, of Third Wave weaponry diffuse throughout the world, spurred on by the process of civilianization.

THE MARRIAGE OF PEACE AND WAR

Until recently the main U.S. defense companies segregated their military business from their other, civilian activities. Today, says Hank Hayes, president of Texas Instrument's defense and electronics group, "if we had to write a vision of what we would like to see happen [it would be] for defense and commercial to merge so that you could in fact run military and commercial manufactured products right on the same manufacturing line."

At another level, the technologies themselves are blending. An indication of the long-term direction of change came in Washington in 1990 when the Department of Commerce and the Department of Defense, normally rivals for political clout, each independently came up with a list of the most important emerging technologies. Which technologies were most needed to prod economic growth? Which were needed for their military potential? Except for a few, the two lists looked remarkably alike.

Similarly, the French government, actively promoting a fusion between commercial and military space efforts, has identified key technologies in which, as *Defense News* reports, "the distinction between a military and a civilian space application is all but lost." The U.S. Army, meanwhile, in a recent white paper, suggested that it could get more for its dollars by reducing, where possible, specialized military specifications, and relying on commercial standards instead.

FAXING THE PARTS

What we may well see, therefore, is the eventual disappearance of most special-purpose military technology companies or their fusion with nonmilitary commercial organizations. The old military-industrial complex will melt into the new civilian-military complex.

This coming fusion casts a sharply different light on present efforts at conversion. As C. Michael Armstrong, chairman of Hughes Aircraft, one of the largest U.S. defense manufacturers, proudly explains, "We can turn military air defense into civil air traffic control. Sensors that warn of chemical warfare can be used to detect pollutants; signal processing can yield digital telephone systems; cruise control radars and infrared night vision can lead to automotive safety systems." He neglected to note that the opposite may be equally true — and not just for Hughes.

Researcher Carol D. Campbell, in looking for commercial markets for Hughes, concluded that its artificial intelligence–based technology for pattern recognition, initially designed for missile targeting, could also be used to recognize handwriting — something useful for the U.S. Postal Service. "If our system can tell a B-1 from an F-16 miles away," she explained to *Business Week*, "it can tell an A from a B or a 6 from a 9."

But Hughes is not the world's only designer of pattern-recognition software, and if, say, Pakistan were to come up with handwriting-recognition technology for its postal service, couldn't that be adapted to missile guidance, too?

In Russia, the Chief Directorate of Ammunition and Special Chemistry is proud of its work adapting satellite sensors originally designed to spot American missiles to the task of locating forest fires instead. Does that mean that sensors produced by Russia or anyone else for spotting forest fires might just as easily be converted to spotting missiles?

Or look at "rapid prototyping" technology. Baxter Healthcare is a medical technology firm that has used this method to make quick, customized models of new intravenous solution equipment. Baxter's peaceful purpose is to help its marketing people and to cut down on engineering development time. But intravenous devices are not the only things that can be made with this technology.

Second Wave armies depend on pre-positioned supplies or a gigantic logistics tail to provide, say, spare helicopter parts. Third Wave armies, relying on advanced computing and "rapid prototyping," will before long be able to make many needed items on the spot. The technology can build objects of any desired shape out of metal, paper, plastic, or ceramics, according to instructions transmitted from data bases thousands of miles away. "It is now possible," reports the *New York Times*, "to, in effect, fax parts to remote sites." This and similar technologies will speed and simplify the projection of military power, reducing the need for permanent foreign bases or supply depots.

For approximately $11,000, Light Machines Corporation of Manchester, New Hampshire, markets a desktop lathe that can cut prototypes out of aluminum, steel, brass, plastic, or wax and can be set up to receive instructions from a remote site.

In short, new knowledge-intensive goods, services, and components technologies are spewing into the global market faster than anyone can track and drastically altering the rules of both peace and war. They will also change the global distribution of arms. If key components of tomorrow's weapons come from civilian production, what countries will be the most important arms suppliers? Those with smokestack factories still stamping out military-specific goods? Or those whose civilian economy is most advanced and best at exports. Until now the Japanese constitution has banned Japanese firms from selling arms. But what about ordinary, innocent civilian goods, software, or services that can be converted and configured for military use? Crucial elements of tomorrow's arsenals could come from the most surprising sources.

When we take account of civilianization, therefore, against the background of today's news, filled as it is with secession movements demanding nationhood, genocidal "ethnic cleansers," criminal syndicates, mercenary forces, have-gun-will-travel fanatics, and various two-bit strongmen and Saddam-clones, the emergent global system takes on a more and more sinister look. It is a world seething with potential violence in which anyone's military edge, including even that of the United States, could be offset or neutralized in unexpected ways. In war and wealth-creation alike, knowledge-intensivity can give power but just as quickly take it away.

In our last book, *Powershift*, we wrote: "By definition, both force and wealth are the property of the strong and the rich. It is the truly revolutionary characteristic of knowledge that it can be grasped by the weak and poor as well. Knowledge is the most democratic source of power."

It may also be the most dangerous. Like the six-shooter in the wild West, it could prove to be the Great Equalizer. The result, however, might not be equality — or democracy. As we see next, it could turn out to be radioactivity, instead. . . .

20

THE GENIE
UNLEASHED

ON A BRIGHT SPRING MORNING recently eight of
us met to decide whether or not to drop a nuclear bomb on North
Korea.

Around an octagonal table littered with Styrofoam coffee cups,
papers, and an open attaché case, we read hastily through the latest
horrifying reports. An attempted coup had just been bloodily suppressed in Pyongyang, North Korea's capital. Its million-plus army
seemed to have split into two factions. Troops were on the move in
the city. Armored units were also streaming across the border toward
Seoul, the capital of South Korea. SCUD missiles, launched from the
North, were hitting targets in the South. American bases there appeared to be under attack from North Korean commando units.

North Korea, we knew, had been building intermediate-range missiles and working on nuclear bombs for years, despite protests from
many countries. Now, with its government apparently tottering,
North Korea did what the world had long feared.

At precisely 9:26 A.M. two North Korean nuclear bombs exploded
over an area where South Korean armor was massing for defense.
Four more nuclear blasts followed three minutes later. In half an hour,
South Korean forces were being attacked by artillery-fired chemical
weapons as well. The Second Korean War had begun with a nuclear
bang.

The task facing our team — and two others — was to put practical
options on the desk of the president of the United States. We had fifty

minutes. The United States was historically committed to the defense of South Korea. Now it faced the question everyone had hoped to avoid: should it respond in kind to North Korea's use of nukes?

At our worktable, a sharp-tongued blond woman pushed for instant retaliation in kind. She was flanked on one side by a slender dark woman who remained grimly silent throughout, and, on the other, by a similarly laconic man with a carefully trimmed gray mustache. All three were from the CIA. A fourth man in blue blazer, regimental tie, and gray flannel slacks urged caution. He was ex-CIA. A stocky, curly-haired man from the Office of the Secretary of Defense broke each suggestion down and pointed out its drawbacks. A cherubic, stripe-shirted nuclear researcher from a leading think tank pushed for non-nuclear options. He was countered by a young academic from Berkeley who believed that hitting 'em fast and hard right from the start would save lives in the end. One of the authors completed the group at our table. Two other tables were ringed by military and intelligence officers, political analysts, decision theorists, and other specialists, all leafing anxiously through the briefing papers and, like us, raising a firestorm of questions.

Who's really in charge in North Korea? Which faction? What do they really want? Who ordered the use of nukes? Do any diplomatic options remain? Should the United States use only conventional forces at first and issue a warning to the North that *further* nuclear use would bring retaliation in kind? Or has the time for warnings passed? If nukes are used, what kind? And how should they be delivered? Ground burst? (No. Too many innocent casualties.) Bombers? Cruise missiles? ICBMs? (No. ICBMs would frighten the Russians and the Chinese.) Should all military targets be hit — or just one? Should the leadership's command bunker be targeted? Minutes raced by. We were already past deadline.... Do we go nuclear or not?

Luckily, no one ever had to make that agonizing decision. The Second Korean War was a fiction — a scenario. The entire exercise was a think-tank game — more accurately a simulation — designed to educate us about potential nuclear crises. It had previously been played by other teams at NATO headquarters in Brussels, as well as by nuclear specialists in the Ukraine and Kazakhstan, two nuclear-armed former Soviet republics.

By the time our game ended, we had looked not merely at what could happen, but also at steps that might be taken in advance to avert such a crisis altogether. But the real nuclear game, of course, is not over. In fact, it is becoming more ominous every day. For that game, like war itself, is being transformed by the arrival of Third Wave civilization and its knowledge-based technologies.

THE DEADLY ANTITHESIS

Nukes, it is worth remembering, did not arise in agrarian societies and were not a part of the First Wave war-form. They came into being in the very last phase of ascendant industrialism. They are the culmination of the search for efficient mass destruction that paralleled the search for efficient mass production. Designed to produce indiscriminate death, they are, in fact, the ultimate military expression of Second Wave civilization.

Today's most advanced weapons are their opposite. They are intended, as we've seen, to de-massify, rather than massify, destruction. But even as Third Wave armies hurry to develop damage-limiting precision weapons and casualty-limiting nonlethal weapons, poorer countries like North Korea, still on the road to Second Wave industrial development, are racing to build, buy, borrow, or burgle the most indiscriminate agents of mass lethality ever created, chemical and biological as well as atomic. Once more we are reminded that the rise of a new war-form in no way precludes the use of earlier war-forms — including their most virulent weapons.

THE NEXT CHERNOBYL

Throughout most of the Cold War only a handful of nations were members of the so-called "nuclear club." The United States and the Soviet Union were charter members. Britain, France, and later China were admitted to "membership."

The sudden split-up of the Soviet Union left the newly independent Kazakhstan, Belorus, and Ukraine with 2,400 nuclear warheads and 360 intercontinental ballistic missiles on their hands. Tortuous negotiations led to agreement that, over a seven-year period, these countries would destroy their strategic weapons or ship them to Russia to be dismantled. Soon, however, Ukraine balked, demanding money for the uranium or plutonium in the warheads. Others hemmed and hawed. The United States was slow delivering promised funds to speed the process. As a result, the task of shipping and dismantling has barely begun.

According to the Russian newspaper *Izvestia*, facilities and maintenance at the Ukrainian missile silos are so poor that another Chernobyl is looming. Workers are exposed to twice the allowable levels of radiation, and security systems have been broken at twenty weapons sites. Meanwhile the Ukrainian minister of the environment has charged that Russia, which is supposed to service and maintain the Ukrainian warheads, has refused to do so until Ukraine admits that they are Russian property — which the Ukrainians refuse to do.

These giant nuclear-tipped ICBMs thus remain targeted at the United States. Some, in Kazakhstan, may be aimed toward China as well. It is not even clear any longer who has or has not cracked their control codes, and therefore which country, if any, is capable of firing them independently.

PADLOCKS AND PERSHINGS

The problem of "small" or tactical nukes is even worse. While tactical nukes cannot "blow up the world," a hailstorm of them could theoretically hit ten or more cities at a time. Individual tacticals can turn as much as a square kilometer of the earth, and everyone on it, into radioactive glass. They can be as small as a few inches in diameter and a foot and a half or two feet long. Many are artillery shells. And at least 25,000–30,000 of these weapons are now in existence.

The United States has withdrawn its tactical nukes from Germany and South Korea. Because the former Soviet republics have by agreement shipped theirs back to Russia, some 15,000 such warheads are now supposedly in Russia. Many more may be squirreled away, however, either undelivered or uncounted in official tallies. Some of these weapons, says one of the Pentagon's top experts, "were old, primitive systems that had no safety devices built in. They might have a padlock cover on them. They're all flavors and they're all over this massive empire. Are they all back in Russia? Statistically, who knows?"

So great is the uncertainty that after the United States destroyed medium-range nuclear missiles in Europe, in accordance with the Intermediate Nuclear Forces Treaty, the American army was, in the words of a Pentagon nuclear specialist, "shocked to find it had another Pershing launcher . . . they hadn't counted. We thought we had blown them up. Then, oh, God, we found another one!" And nuclear-tipped Pershings were far easier to count and identify than the smaller and far more numerous tactical weapons.

In supposedly "safe" Russia today these "small" weapons are

stored in totally inadequate facilities. Says a member of parliament and former Soviet cosmonaut Vitaly Sevastyanov, "The existing depots are chockful of warheads and some are even stored in rail cars." The Russians lack the technical staff, the structures, and above all the money needed to safeguard these weapons.

Governments, criminal syndicates, and terrorist groups around the world are itching to lay hands on even a few of these weapons. The Russian military, in turn, including units supposedly guarding these weapons, are poorly paid, poorly sheltered, and not above corruption. Russian officers have already peddled other weapons to illegal buyers in under-the-table transactions.

In a nightmarish scenario described to us by a Pentagon specialist, a corrupt Russian colonel sells a warhead to a revolutionary terrorist gang based in, say, Iran. When the United States or the UN demands to know what happened, both the Russian and the Iranian governments deny knowledge. And both, in this case, might be telling the truth. Yet one or both might, in turn, be disbelieved. No one knows what mistaken retaliation might ensue.

There is, after all, plenty of reason to disbelieve both (indeed all) governments in these matters. The Iranians may well be lying when they insist that all their nuclear activity has peaceful purposes. Iraq and North Korea said exactly the same thing. According to intelligence sources, Iran has built a hidden network of nuclear research centers. And like Iraq before it, Iran has bamboozled International Atomic Energy Agency (IAEA) inspectors. When they asked to visit the Moallem Kalayah site near Teheran, they were taken to another village with the same name.

According to the People's Mujahedeen, a leading Iranian opposition group, Iran has actually succeeded in buying four nuclear warheads from Kazakhstan. When the authors met with Kazakhstan president Nursultan Nazarbayev in Alma Ata in December 1992 and pointedly asked him about this report, he labeled it mere rumor. The fact is that no one — perhaps not even presidents and their cabinet ministers — know the full truth.

Who should one believe? The interior minister of Azerbaijan, speaking in Baku during the height of its war with Armenia, boasted that it already had acquired six atomic weapons. He may have been bluffing. Or he may not have been. And the world hardly noticed when the prime minister of tiny South Ossetia, an autonomous region in Georgia, threatened to use nuclear weapons belonging to the for-

mer Soviet Union against Georgian paramilitary forces. No one is
sure any longer exactly who is and who is not a member of the once-
exclusive "nuclear club."

BAMBOOZLED INSPECTORS

So long as nukes were the property of big, powerful, and stable regimes, the Second Wave approach to proliferation problems in the
world was relatively simple. Over the years a patchwork of treaties
and agencies were created to police the possible proliferators. The
Nuclear Non-Proliferation Treaty (NPT) and the IAEA were supposed to block the spread of nukes. A Missile Technology Control
Regime (MTCR) was set up to stem the proliferation of missiles.
Other agreements were designed to prevent the diffusion of chemical
and biological weapons. But these instruments are feeble at best.

The NPT has often been hailed as "the most widely adhered to
arms control treaty in history" because 140 parties have signed it. But
countries "adhere" to NPT in direct proportion to its toothlessness.
Nuclear bombs are made from plutonium or from highly enriched
uranium. Of the 3,000 tons of HEU now floating around the world,
only thirty tons — a mere one percent — are actually subject to
IAEA policing. Of the 1,000 tons of plutonium known to exist today,
less than a third is even theoretically under international safeguard.
Moreover, the IAEA's primary task has been to arrange for IAEA
inspectors to visit declared civilian nuclear energy plants to make sure
their uranium or plutonium is not diverted for bomb production. But
this is not the main problem anymore. As both Iraq and North Korea
have shown, the greater problem lies in "undeclared" or secret plants.
And countries can now get these materials in other ways.

Since the end of the Gulf War the public has grown accustomed to
seeing televised pictures of large IAEA teams flying boldly into
Baghdad. But the IAEA is no more than a gnat on the hide of a
radioactive rhinocerous.

In November 1990, three months after Saddam had already invaded Kuwait, the IAEA sent a team to Baghdad. Shown only what
the dictator wanted them to see, they, needless to say, gave Iraq a
clean bill of health. One had to read the fine print to learn that the
team consisted of exactly two (2) inspectors who were supposed to
verify the peaceful intent of what turned out to be one of the world's
most aggressive, multifaceted bomb-building projects.

Even after the Gulf War, when teams of IAEA inspectors went into Iraq under UN Security Council mandate, the agency's performance was appalling. Its chief inspector, Maurizio Zifferero, in September 1992 reportedly announced that Iraq's bomb program was "at zero." Yet by early 1993 his inspectors discovered still another mass of equipment that clearly contradicted his premature, perhaps self-deceptive optimism.

CHICKEN-CHECKERS

Prior to the Gulf War the IAEA used the equivalent of only 42 full-time inspectors to check on 1,000 declared nuclear energy plants around the world. By contrast, the United States fields 7,200 full-time inspectors to check on salmonella or psittacosis in its meat and poultry — 171 for each and every inspector sent by the world community to check the spread of the world's nuclear disease. In effect, it spends two and a half times more each year to make sure its chicken and beef are OK than the IAEA spends to guarantee nuclear safety on the globe ($473 million versus $179 million).

Even the post-Gulf strengthening of the NPT, and the new support given it by the UN Security Council, leave it a laughing matter to violators of the treaty and to nonsignatories. The gnat is still a gnat.

PORNOGRAPHY AND HEROIN

With all the world's satellites, spies, and sensors, one would think finding nukes or nuclear facilities would by now be relatively simple. But as the Iraq case proves, that is hardly the case. Shielded in enough lead and paraffin and lowered deep enough into the earth, a nuclear warhead can be quite undetectable. The technologies of detection have not caught up with even the primitive forms of concealment.

At the same time, the spread of peacetime nuclear energy plants has increased the world output of wastes from which warheads can be built. Channels of international trade are fast-multiplying, too — among them channels for smuggling nuclear materials, machines, and/or warheads. And, in the words of the *Moscow Times*, "Russia's borders have become sieves through which every type of good in every state of matter — liquid, solid, and gas — is finding its way out."

When the authors met in Moscow with the Russian Minister of Atomic Energy, Viktor Mikhailov, we heard words of syrupy reassurance. Nevertheless, when 3.3 pounds of highly enriched uranium disappeared from an institute in Podolsk, near Moscow, the ministry's own head of internal security, Alexander F. Mokhov, said, "The thefts were carried out by people directly linked to the technical processes, who know them superbly. They knew how to steal, bit by bit, so it cannot be detected." Less-sophisticated would-be smugglers, with less-enriched material, have been captured by police in Austria, Belarus, and Germany, where police report over one hundred cases of illegal movements of nuclear materials.

The radically new situation in the 1990s confirmed nuclear strategist Thomas Schelling's warning, in 1975, that "we will not be able to regulate nuclear weapons around the world in 1999 any better than we can control the Saturday-night special, heroin or pornography today."

WALL STREET AND WARLORDS

All this leads some pessimists to doubt that nuclear arms can be controlled at all. Few match the gloom of Carl Builder, a strategic analyst at the RAND Corporation. Builder's pessimism is regarded as extreme by many of his colleagues, but as the first director of nuclear safeguards for the U.S. Nuclear Regulatory Commission, he can hardly be dismissed. At one time Builder was totally responsible for the security of all nuclear materials in civilian hands in the United States, some of it bomb-grade stuff.

The main nuclear problems of the future, he believes, will arise not from nation-states at all, but from those we called "global gladiators" in our book *Powershift*. These are terror organizations, religious movements, corporations, and other nonnational forces — many of whom, he says, could gain access to nuclear weaponry.

Listening to him one imagines the Irish Republican Army announcing that it has acquired its own nuclear bomb. A call to the BBC warns that "if British troops do not evacuate Northern Ireland within seventy-two hours, a nuclear device will . . ." The bumblers who devastated parts of New York's World Trade Center might have obliterated Wall Street had someone cleverer supplied them with a tactical nuke. Someday, Builder believes, even outfits like the Medellin cocaine cartel may be able to build their own atomic weapons.

According to a report in *The Economist*, "There have already been more than 50 attempts to extort money from America with nuclear threats, some frighteningly credible." Worse yet, to the current list of possible threats an additional one, largely overlooked, now has to be added. Not only governments, terrorists, and drug barons, but warlords may now be searching for nuclear arms.

There are, often ignored by the arms-control community, private armies in many parts of the world under the control of local business-cum-political thugs. The equivalent of warlords can be found from the Philippines to Somalia and the Caucasus, wherever central government control is weak. More and more of these private armies are springing up as the national forces of the old Soviet Union disintegrate. Moreover, there are reasons to believe that mafia-like business groups in Russia today feed, house, clothe, and control whole units of the former Red Army. In short, private armies, mercenaries, and First Wave warlordism are all making a comeback. The idea of nuclear weapons under the control of these local generalissimos should send a shudder down our collective spine.

Builder's proliferation scenario, however, forces us to confront the extreme. Like gunpowder, he says, "Nuclear weapons are going to diffuse. . . . I'm going to go even further and say, even if not in my lifetime, perhaps, but in the foreseeable future, [that they] are going to proliferate down to individuals. It will be possible for an individual to make a nuclear device from materials which are in commerce."

Mafia families, Branch Davidian cultists, archaeo-Trotskyite groupuscules, Sendero Luminoso Maoists, Somalian or Southeast Asian warlords, Serbian Nazis, and even, perhaps, individual loonies could hold whole nations at ransom. Worse yet, Builder believes, "An opponent cannot be deterred by the threat of nuclear weapons if that opponent has no definable society to threaten." Thus, he says, a "terrifying asymmetry" looms ahead.

THE BROKEN DAM

The dam that is still supposed to hold back the flood of mass-destruction weaponry depends not merely on ineffective treaties and inspection systems, but on a patchwork of export controls. Enacted by various governments, these supposedly prevent the transfer of components and materials needed for mass-destruction weapons. But within the United States alone, says Diana Edensword of the Wisconsin Project on Nuclear Arms Control, one finds a tangle of "uncoordinated and overlapped export agencies."

At a global level, the lack of coordination is even more apparent. Each country applies different standards and definitions — different lists of what products or technologies should be nonexportable. Enforcement levels vary constantly. And if anti-nuke programs are a mess, there is even less coherence or coordination among programs focused on missiles, chemical weapons, or biological-warfare toxins. In short, there simply is no effective *system* to stop the spread of Second Wave weapons of mass destruction.

When we place such facts side by side, we discover a revolutionary situation never anticipated by official arms-control agencies, peace groups, and nonproliferation experts.

Even if we totally ignore the mounting threat from nongovernmental groups, and focus on nation-states alone, we can conclude that approximately twenty countries are either in or knocking at the door of the Nuclear Club. Indeed, according to former Ambassador Richard Burt, who helped negotiate nuclear build-down agreements between the United States and Russia, some fifty to sixty countries can acquire these weapons. And if, instead of a nuclear club we imagine a Mass Destruction Club, with a broader membership that includes countries with chemical and biological weapon capabilities or ambitions, that number would leap upward. We may be looking at a world in which a third to a half of all countries have some hideous weapons of mass murder tucked away in their arsenals.

PULVERIZED PREMISES

Asked what went wrong, how the genie escaped out of the bottle, most experts blame the breakup of the Cold War world. But that answer is inadequate.

It is the coming of the Third Wave — with its knowledge-intensive technologies, its corrosive impact on nations and borders, its information and communications explosion, its globalization of finance and trade — that has pulverized the premises on which arms-control programs have until now been based.

Second Wave efforts at preventing the spread of mass-destructive weapons were based on ten key assumptions:

1. The new weapons could be monopolized by a few strong nations.
2. Nations seeking such arms would have to produce their own.

3. Small nations, in general, lacked the necessary resources.

4. Only a few weapons or types would meet the definition of weapons of mass destruction.

5. These weapons depended on a handful of raw materials that were monitorable and controllable.

6. They also depended on a few specific, identifiable technologies whose spread could also be watched and controlled.

7. The actual number of "secrets" needed to prevent proliferation would also be small in number.

8. Regulatory agencies like the IAEA could collect and disseminate information for use by the world nuclear industry without revealing knowledge that would help arms proliferators.

9. Existing nations would remain stable and not break apart.

10. Nation-states were the only possible proliferators.

Today every one of these assumptions is demonstrably false. With the rise of the Third Wave, the Second Wave threat of mass destruction has been totally transformed.

FLEX-TECHS

One of the relative handful of people who worry day in and day out about this revolution is a ruddy-faced navy intellectual named Larry Seaquist. As intellectuals go, he has had an unusual career.

Son of a farmer and his wife in eastern Idaho, Seaquist grew up with a sense of adventure fostered by copies of *National Geographic* magazine. Through luck and initiative, he landed a job with a private company doing meteorological readings in the Arctic in connection with the DEW line — the chain of distant early-warning radar stations that ran from Greenland across Canada to Alaska along the Seventieth parallel, 200 miles north of the Arctic Circle. While wintering there, he heard that the U.S. Weather Bureau was looking for volunteers to go to the South Pole with an Argentinian expedition. After a stint at language school to learn Spanish, he flew out on the first Argentine flight to the pole, and spent fourteen months on the Antarctic ice. By the time he was twenty-three, he had been at both ends of the earth.

Seaquist later joined the U.S. Navy, rose to command the famed battleship the USS *Iowa* — the ship that suffered a devastating accidental explosion some years after Seaquist's departure. After his commands at sea, Seaquist became a top-level strategist for the navy, moved to Washington with his playwright wife, Carla, and went to work for the Joint Chiefs of Staff in the Pentagon. He ultimately joined the Office of the Secretary of Defense as Special Coordinator in a small policy team responsible for rethinking the unthinkable.

One result of its work is a radical redefinition of the entire proliferation threat. Proliferation for Seaquist is defined as "the destabilizing spread, especially to countries of concern in key regions, of a wide array of dangerous military capabilities, supporting capabilities, allied technologies, and/or know-how." This definition itself represents a sharp break with the past, both deepening and broadening the meaning of the term.

Until now "nonproliferation" policies focused narrowly on weapons, delivery systems, and certain space systems. The new concept is called "counter-proliferation" and it deals with "capabilities" in general, which include technologies and knowledge. Thus, in assessing a country's policy toward weapons of mass destruction, it looks beyond the nation's hardware to its military doctrine, its training, and other intangibles.

It especially focuses attention on Third Wave knowledge-driven technology — the new "flex-techs" capable of constantly changing their output to meet varying needs. They provide the basis for the process of civilianization described in the last chapter, and they change all the proliferation equations.

As Seaquist explains, "The proliferation of advanced manufacturing machinery around the world is very important. Numerically controlled machinery is now in many Third World countries.... A pharmaceutical plant that they need ... has the inherent capability to manufacture biological weapons. Numerically controlled machinery that manufactures good quality automobiles in the Third World can also manufacture good quality rockets."

The rapid spread of these quintessentially Third Wave machines powerfully shifts military balances — and threatens to deprive the United States of its predominance. Except for a superior ability to integrate advanced technologies and military forces, he contends, the United States has "no technological monopoly in virtually anything."

In fact, Seaquist says, "I've never found anyone to respond to my challenge to name three technologies which are under the exclusive control of the U.S. military. There's nothing left. We used to, if it was something important, keep it from the Russians. Or, if they developed it, they'd try to keep it from us. We were on parallel tracks and everyone was left behind. . . . Not now."

Behind the actual hardware, of course, lies the ultimate intangible: know-how. We are seeing a rapid, worldwide de-monopolization of all kinds of information. Even doctors can no longer control the flood of medical knowledge into society through the media and other channels. This process of de-monopolization, driven by commercial and other necessities, has broad democratic implications under some circumstances — and destabilizing military implications under others.

FREEDOM OF INFORMATION (FOR BOMB-BUILDERS)

Much of the know-how to produce nuclear arms (maybe not the most powerful types, but powerful enough) has been disseminated to just about anyone who wants it, terrorist, maniacal crank, or pariah nation. Want to build a bomb? Got a PC? Log onto the IAEA's friendly data bank, the International Nuclear Information System, for a headstart. Go to the vast, open literature available in technical libraries. Buy an underground nuclear cookbook entitled *Basement Nukes*, a copy of which we scan as we write. This pamphlet, too, is openly on sale if one knows where to look. Says Michael Golay, a professor of nuclear engineering at the Massachusetts Institute of Technology, "What's classified today is how to build a good weapon, not how to build a weapon."

But it isn't just the spread of flex-techs or the leak of "secrets" that has created today's dangerous new reality. The RAND Corporation's Carl Builder points out that "military programs will have less effect on the nature of nuclear deterrence than the political and social changes now being wrought by the information era."

For example, "The flow of information into or out of a nation can no longer be effectively controlled by the state; information is everywhere and accessible. To participate in the burgeoning economic benefits of world commerce means adopting practices that undermine state control. . . .

"The roots of national power in the industrial era were thought to lie in natural resources and plant investment. . . . In the information era" — that is, the Third Wave era — "those roots now appear to be in the free access to information."

This is the deeper force transforming the threat environment and the proliferation problem. For this, Builder says, is why "the information necessary for development of nuclear weapons will inevitably spread beyond the control of the nation-state" and why "commerce will make nuclear materials or the means for their production (or recovery) increasingly available world-wide."

What goes for nuclear weapons, is equally applicable to other weapons as well. And that compels those who wish a more peaceful world to recognize the dilemma of the twenty-first century.

We will either have to slow the development and diffusion of new knowledge — which is immoral if not impossible — to prevent wars of mass destruction. Or we will have to accelerate the collection, organization, and generation of new knowledge, channeling it into the pursuit of peace. Knowledge is what the anti-wars of tomorrow will be about.

The new dangers the world faces from the civilianization and proliferation of weapons, however, are set against an even broader set of threats to peace — new dangers in a new world. To understand these, we must pass beyond the Zone of Illusion.

21

THE ZONE
OF ILLUSION

ONE of the lingering aftereffects of the collective ecstasy that gripped the world after the Berlin Wall fell is a conviction that even if wars proliferate in the years ahead, they will barely touch the high-tech democracies. The unpleasantness will be confined to local or regional conflicts, mainly among poor, dark-skinned people in remote places. Not even the outbreak of war and genocide in the Balkans dented the complacence of West Europeans on whose doorstep the blood was spilled.

The potential for many smallish, "niche" wars in First Wave and Second Wave regions is, indeed, growing. But this should not lead us to the conclusion that major powers will remain safely at peace. Just because the danger of escalation to an all-out U.S.-Soviet nuclear exchange has diminished does not mean the danger of escalation itself has disappeared. The widening proliferation of weapons of mass destruction, the growing application of civilian technology to military purposes, the weakness of anti- and counter-proliferation regimes all point to the possibility of "small" wars getting bigger and nastier, and spreading across borders — including the borders of the so-called Zone of Peace in which the high-tech powers dwell, and in which war is supposedly inconceiveable.

It is increasingly difficult to cordon off parts of the global system from disruptions or destruction in other parts. Masses of immigrants spill across borders, sometimes bringing their hatreds, political movements, and terrorist organizations with them. Abuse of an ethnic or religious minority in one state triggers cross-border repercussions in another.

Pollution and disasters respect no boundaries and trigger political unrest. Any or all of these could suck major, high-tech economies into conflicts they do not want but do not know how to limit or prevent.

This is not the place to catalog all the bloody conflicts currently raging around the planet, many with significant risks of escalation and contagion. We may similarly pass over the dangers posed by an unstable, nuclear-armed Russia.

We can even, perhaps, continue to ignore the fact that the Asia Pacific area, containing the world's hottest, most important economy, is increasingly unstable, both politically and militarily.

Though few seem to have noticed it, this region, the core of the entire global economy, is more tightly ringed with nuclear weapons than any other part of the world. (The perimeter of the area, from Kazakhstan, India, and Pakistan to Russia, China, and North Korea, consists of nuclear and near-nuclear countries, many of them politically volatile.)

India is rent by religious fanaticism and is fighting several different armed insurgencies at the same time. China's political future remains a question mark, even as its air force extends its reach with Russian-built Sukhoi fighters and aerial refueling capability and its navy hungers for an aircraft carrier.

Taiwan responds to China's moves by buying 150 F-16 fighters from the United States and fifty to sixty Mirage jets from France. Other arms races proliferate throughout the region. Watching all this, Japan — once the world's most fervently anti-nuclear nation — suddenly announces it will not support indefinite extension of the Non-Proliferation Treaty. It is a clear message that it no longer rules out building nuclear weapons of its own. Yet this is the moment when isolationists in the United States, against the fervently expressed wishes of most Asian nations, contemplate cutting costs by shrinking its military presence in the Western Pacific — threatening, in effect, to yank out or weaken the region's stabilizer of last resort.

But even if we brush these and other looming regional troubles aside, we are left with a set of emerging *generic* problems, any one of which could explode in our faces in the decade or two to come. These global "generics" compel us to question the theory that the great powers, or even the great democracies, inhabit a zone of peace in which war is unthinkable. Alas, the zone-of-peace notion needs to be interred alongside the corpse of geo-economics.

Consider the possibilities.

A MONEY MELTDOWN

Imagine a real, worldwide meltdown of the money system. So far the major economies have faced only a mild recession since the end of the Cold War. What happens to the unthinkability of war in the so-called zone of peace if the world plunges into a real market-crushing global depression? A depression brought on, perhaps, by protectionist trade wars, managed trade, and other forms of "geo-economic" competition?

Today's financial system is, in fact, extremely vulnerable because it is in the process of restructuring itself to service a fast-globalizing Third Wave economy. In liberalizing flows of capital across national divisions, myopic politicians and financial leaders have dismantled many of the fail-safe devices and brakes that once might have limited the effects of a serious collapse to a single nation. They have done little to replace these safeguards.

The last relatively minor dip in the world economy coincided with neo-Nazi terror in Germany and a Los Angeles set ablaze. Even Japan, that most orderly of societies, felt the first tremors of social unrest as its "bubble economy" burst. What would happen to peace and stability in the supposedly war-proof zone if the world financial system really crashed — a prospect that cannot be ruled out?

BOUNDARY BREAKAGE

Western media today describe outbreaks of ethnic conflict in the Balkans and the Caucasus as a function of "backwardness." We may soon find, however, that border-busting is not just a result of "tribalism" or "primitive ethnicism."

Two other forces are challenging national borders. The emergent Third Wave economy, based on knowledge-intensive manufacture and services, increasingly ignores existing national boundaries. As we already know, large companies form cross-border alliances. Markets, capital flows, research, manufacture — all are reaching out beyond national limits. But this highly publicized "globalization" is only one side of the story.

New technologies are simultaneously driving down the cost of certain products and services to the point at which they no longer need national markets to sustain them. No one has to send snapshots to Kodak in Rochester, New York, for processing any more. It can be done faster and cheaper on the nearest street corner, using small-scale, inexpensive, decentralized technology. Such small, cheap, miniaturized technologies are spreading rapidly.

Enough such decentralized technologies could in time change the entire balance between national and regional economies. They make the latter more viable, thus strengthening the hand of border-breaching separatist movements. Simultaneously, the growing number of television channels, whether over-the-air, satellite-based, or cable, point to more localized programming in more languages, from Gaelic to Provençal, providing cultural support for the technical and economic forces described here.

Europe is already awash in secessionist, autonomist, or regionalist groups from northern Italy to Spain and Scotland. They seek to redraw its political maps and move power downward from the nation-state, even as Brussels and the European Community drain power away from the nations and move it upward.

Twin changes, therefore, one from above and the other from below, are cutting the ground out from under the rationale for national markets — and the borders they justify.

These pincer pressures pit inflamed nationalists, regionalists, and localists, including some who aspire to "ethnically cleanse" their turf, against the more cosmopolitan Europeanists — hardly a recipe for continued stability in that "zone of peace."

No border seems more permanent than that which exists between the United States and Canada. But many Quebecois already believe they can flourish economically without the rest of Canada. Should Quebec, after decades of struggle, ever secede from Canada, British Columbia and Alberta might soon after seek admission into the United States. Another scenario (certainly implausible but not impossible) pictures formation of a new political entity — whether called a nation-state or not — uniting these western provinces of Canada with the American states of Washington, Oregon, and maybe Alaska.

Such a federation or confederation could start life with vast resources, including Alaskan oil; Albertan natural gas and wheat; Washington State's nuclear, aerospace, and software industries; Oregon's timber and high-tech industries; giant ports and transport facilities serving the Asia Pacific trade; plus a highly educated work force. It could, at least in theory, become an instant economic giant with a massive trade surplus — a key player in the world economy.

Some forecasters see a future world not with today's 150–200 states, but with hundreds, even thousands of mini-states, city-states, regions, and noncontiguous political entities. The decades to come will see even stranger possibilities emerge as existing national boundaries lose their legitimacy and the border-busters go to work in the very heart of the zone of peace.

MEDIA GOVERNMENT

The idea that democracies don't fight one another also presupposes that they stay democratic. In Germany, for example, even as we write, many question whether it is safe to make that assumption.

Staying democratic, in turn, takes for granted a degree of political stability or orderly change. Yet many of the nations in the presumed zone of peace are speeding into a turbulent period of political perestroika, or restructuring.

As they shift from muscle-based to mind-based economies, massive layoffs and dislocations accompany the rise of a new political force — a high-skill "cognitariat" that is displacing a low-skill proletariat. As knowledge becomes the central economic resource and electronic networks and media become the critical infrastructure, those in command of knowledge and the means of communication grab for enhanced political power.

One indication of this is the radically enhanced political influence of the media, nowhere more evident than in the American election of 1992, when a single television network, CNN, played a decisive role in the defeat of President George Bush. Only a year earlier the same CNN, with its extensive coverage of the Gulf War, helped push Bush's popularity to extraordinary heights.

Seven months later Republican Bush lost his bid for reelection. Democrat Bill Clinton won — but scored fewer votes than his party's previous candidate, Michael Dukakis, who lost in 1988. Clinton won with this small tally because a third candidate, Ross Perot, siphoned votes away from both major-party candidates, and an intramural fight, led by Pat Buchanan inside the Republican Party, further damaged Bush.

Perot, the Billionaire Politician, was virtually a creature of CNN, having launched his campaign in front of its cameras and having appeared frequently on its channels thereafter. Buchanan, just prior to his political campaign, was, in fact, the on-camera cohost of the daily CNN show *Crossfire*. In no previous political campaign in the United States did the media, let alone a single channel, play so crucial a role.

But the new media do more than change election outcomes. By focusing the camera first on one crisis, then almost overnight on another, the media increasingly set the public agenda and force politicians to deal with a constant flow of crises and controversies. Abortion today. Corruption tomorrow. Taxes next. Then sexual harassment . . . government deficits . . . racial violence . . . disaster relief . . . crime. . . . The effect is to accelerate political life — compelling governments to make decisions about increasingly complex matters at an increasingly faster clip. They become victims, as it were, of future shock.

But what we have seen so far is only the opening barrage in the media's coming drive for political power. Much of the Clinton-Bush-Perot campaign was waged on call-in shows, the early, still-primitive form of media interactivity. Since then radio talk shows, responding instantaneously to government proposals, appointments, and scandals, have begun systematically giving expression to, and even organizing, political dissent. Talk jockeys can deluge Washington with letters, angry phone calls, and soon — no doubt — delegations.

But as suggested earlier, all this is still foreplay. Television sets of the future will simplify and universalize interactivity, reducing the power of the one-way communications on which politicians and governments have depended since the origins of the mass media in the early part of the industrial revolution.

Today's slow-moving congresses, parliaments, and courts are products of the First Wave. Today's giant ministries and governmental bureaucracies are largely products of the Second Wave. Tomorrow's media — from cable television to direct-broadcast satellite to computer networks and other systems — are products of the Third Wave. The people running them are about to challenge preexisting political elites — and thereby transform political struggle.

In every modern democracy incessant political warfare has, until now, been waged between politicians and bureaucrats. This covert struggle for power is often more important than the overt battle between parties of left or right. With rare exception, this is the real nature of political struggle, from Paris and Bonn to Tokyo and Washington.

As the media's political clout increases, however, the old two-way battle becomes a three-way struggle for power, pitting parliamentarians, bureaucrats, and now media people against one another in unstable combinations.

Meanwhile, hurricanes of religious proselytizing, political propaganda, and popular culture will come storming into each country from outside its borders via direct-broadcast satellite and other advanced telecommunications systems, further weakening politicians and bureaucrats alike in the host country. Trans-border digital

networks with names like GreenNet, GlasNet, PeaceNet, and Al-
ternex already link political activists in ninety-two countries from
Tanzania and Thailand to the United States and Uruguay. Neo-Nazis
have their own nets. In tomorrow's "mediatized" political systems,
consensus will be harder and harder to manufacture from the top.

As the power struggle is played out between elected politicians,
appointed bureaucrats, and media representatives who are neither
elected nor appointed, the military leaders of democratic states will
find themselves trapped in a double bind. The democratic principle of
civilian control of the military itself may be endangered. Since military threats and crises can materialize faster than consensus can be
organized, the military may be paralyzed when action is required. Or
it may, conversely, plunge into war without democratic support.

In either case political perestroika promises the exact opposite of
the stability that the zone-of-peace concept takes for granted.

INTERNATIONAL OBSOLESCENCE

Worse yet, the old tools of diplomacy will prove obsolete — along
with the UN and many other international institutions.

Much foolishness has been written about a new, stronger United
Nations. Unless it is dramatically restructured in ways not yet even
under discussion, the UN may well play a less effective and smaller,
not larger, role in world affairs in the decades to come.

This is because the UN remains what it originally was, a club of
nation-states. Yet the flow of world events in the years ahead will be
heavily influenced by *nonnational* players like global business, cross-
border political movements like Greenpeace, religious movements
like Islam, and burgeoning pan-ethnic groups who wish to reorganize
the world along ethnic lines — the Pan-Slavs, for example, or certain
Turks who dream of a new Ottoman Empire that unites Turks and
Turkic speakers from Cyprus in the Mediterranean to Kyrgyzstan on
the Chinese border.

International organizations unable to incorporate, co-opt, en-
feeble, or destroy the new nonnational sources of power will crumble
into irrelevance.

THE MENACE OF INTERDEPENDENCE

One final comforting myth built into the zone-of-peace notion needs
correction — the myth of peaceful interdependence.

Geo-economists and others may argue that military conflict is lessened when nations become more dependent on one another for trade or finance. Take Germany and Britain, they say, old adversaries now at peace. What this example overlooks is that when Germany and Britain went to war against each other in 1914, each was the other's biggest trading partner. History books provide plenty of other examples as well.

More important and even less noticed is the fact that while interdependence may create bonds between nations, it also makes the world far more complicated. Interdependence means that Country A cannot take an action without triggering consequences and reactions in countries B, C, D, and so on. Certain decisions taken in the Japanese Diet can have more impact on the lives of American auto workers or real estate investors than decisions taken in the U.S. Congress — and vice versa. The shift to fiber optics in the United States can, in principle, push down copper prices in Chile and cause political instability in Zambia, whose government revenues depend on copper exports. Environmental regulations in Brazil can change timber prices and the lives of loggers in Malaysia, which, in turn, can shift political relations between its central government and the sultans who rule various regions.

The greater the interdependence, the more countries are involved and the more complex and ramified the consequences. Yet interrelationships are already so tangled and complex that it is nearly impossible for even the brightest politicians and experts to grasp the first- or second-order consequences of their own decisions.

This is another way of saying that, except in the most immediate sense, our decision makers no longer really understand what they are doing. In turn, their ignorance in the face of enormous complexity weakens the links between goals and actions, and increases the level of guesswork. Chance plays a bigger role. Risks of unanticipated consequences skyrocket. Miscalculations multiply.

Interdependence, in short, doesn't necessarily make the world safer. It sometimes does just the opposite.

In brief, every one of the assumptions on which the zone-of-peace theory is based — economic growth, the inviolability of borders, political stability, time for negotiation and consultation, the effectiveness of international organizations and institutions — is now highly dubious.

While they may seem unrelated to one another, every one of the new, more dangerous conditions described here is a direct or indirect consequence of the rise of a new wealth-creation system. These generic problems indicate potentially deadly trouble ahead. Taken together with the civilianization and proliferation of weapons, they point not to an era of geo-economic peace, to a stable new world order, or a democratic zone of peace, but to a growing risk of war, involving not just small or marginal powers but the great powers themselves.

Nor do these exhaust the long-range dangers we face. As we will see next, we also face several challenges of even greater historical scale and scope — any one of which could produce, if not a world war, then something horrifyingly similar.

To reduce these risks, we need to be brutally realistic about the coming transformation of war and anti-war. We need to move out of the zone of illusion.

22

A WORLD TRISECTED

FOR CENTURIES elites have feared and protected themselves against revolts of the poor. The history of both agricultural and industrial societies is punctuated with blood-spattered slave, serf, and worker uprisings. But the Third Wave is accompanied by a startling new development — an increasing risk of revolt by the rich.

When the U.S.S.R. broke apart, the republics most eager to split away were the Baltic states and the Ukraine. Closest to Western Europe, they were also the most affluent and the most industrially developed.

In these Second Wave republics the elites — chiefly Communist Party bureaucrats and industrial managers — felt hamstrung and overtaxed by Moscow. Looking westward, they could see Germany, France, and other nations already moving beyond traditional industrialism toward a Third Wave economy. They hoped to hitch their own economies to the West European rocket.

By contrast, the republics most reluctant to leave the Union were the farthest from Europe, the poorest and most agrarian. In these heavily Muslim First Wave republics, the elites called themselves Communist, but often resembled corrupt feudal barons operating through highly personal family and village networks. They looked to Moscow for protection and handouts. Second Wave and First Wave regions thus pulled in sharply opposed directions.

All sides masked their self-interest in flag-waving ethnic, linguistic, even ecological appeals. Behind the resultant clashes, however, lay sharply opposed economic and political ambitions. When the contrary pulls of the First and Second Wave regional elites became too strong for Gorbachev to reconcile, the great Soviet crack-up ensued.

THE CHINA SYNDROME

An X ray of other large nations reveals similar fault lines based on First, Second, or Third Wave differences.

Take, for example, China, the world's most populous country. Today, out of its 1.2 billion people, as many as 800 million are peasants in the interior, still scrabbling at the soil much as their grandparents did under conditions of wretched poverty. In Guizhou and Anhui the swollen bellies of hungry children are still all too visible amid shacks and other marks of misery. This is First Wave China.

By contrast, China's coastal provinces are among the most rapidly developing in the entire world. In factory-filled Guangdong, gleaming new high rises pierce the sky and entrepreneurs (including ex-Communist functionaries) are plugged into the global economy. Looking nearby, they can see Hong Kong, Taiwan, and Singapore swiftly transforming themselves from Second to Third Wave high-tech economies. The coastal provinces view these three so-called "Tigers" as models for their own development, and are linking their own local economies to them.

The new elites — some engaged in Second Wave enterprises based on cheap labor, others already installing leading-edge Third Wave technologies at a blistering pace — are optimistic, extremely commercial, and aggressively independent. Equipped with faxes, cellular phones, and luxury cars, speaking Cantonese, rather than Mandarin, they are wired into ethnic Chinese communities from Vancouver and Los Angeles to Jakarta, Kuala Lumpur, and Manila. They share more in life-style and self-interest with the Overseas Chinese than with First Wave China on the mainland.

They are already thumbing their collective nose at economic edicts from Beijing's central government. How long before they decide they will no longer tolerate Beijing's political interference and refuse to contribute the funds needed by the central government to improve rural conditions or to put down agrarian unrest? Unless Beijing grants them complete freedom of financial and political action, one can imagine the new elites insisting on independence or some facsimile of it — a step that could tear China apart and trigger civil war.

With enormous investments at stake, Japan, Korea, Taiwan, and other countries might be compelled to take sides — and thus find themselves sucked unwillingly into the conflagration that might follow. This scenario is admittedly speculative, but not impossible. History is dotted with wars and upheavals that looked highly improbable.

THE RICH WANT OUT

India, with a population of 835 million, is the world's second most populous state, and it is developing a similar split among its trisected elites. There, too, a vast peasantry still lives as in centuries past; there, too, we find a large, thriving industrial sector of roughly 100 to 150 million people; and there, too, we find a small, but fast-growing Third Wave sector whose members are plugged into Internet and the world communications grid, working at home on their PCs, exporting software and high-tech products, and living a daily reality radically different from the rest of society.

A glance at MTV blaring out over Indian television screens or a visit to the Lajpat-Rai market in South Delhi makes the cleavage between the sectors clear. There customers haggle with hucksters over the price of satellite dishes, LEDs, signal splitters, video recorders, and other gear needed to plug into the world's Third Wave infostream.

India is already torn by bloody separatist movements based on what appear to be ethno-religious differences. If we look beneath these, however, we may find, as in China and Russia, three opposed elites, each with its own economic and political agenda, tearing the nation apart under the guise of religion or ethnicity.

Brazil's population of 155 million is seething, too. Nearly 40 percent of the work force is still agricultural — much of it barely existing under the most abominable conditions. A large industrial sector and a tiny but growing Third Wave sector make up the rest of Brazil.

Even as masses of First Wave peasants from the Northeast starve, and out-of-control migrations overwhelm Second Wave São Paulo and Rio, Brazil already faces an organized separatist movement in Rio Grande do Sul, an affluent region in the South with an 89 percent literacy rate and a phone in four out of every five homes.

The South produces 76 percent of the country's GDP and is routinely outrepresented in government by the North and Northeast, whose economic contribution, measured in these terms, is only 18 percent. The South, moreover, argues that it is subsidizing the North.

Joking that Brazil would be rich if it simply ended just north of Rio, southerners are no longer laughing. They claim they send 15 percent of their GDP to Brasilia and receive only 9 percent back.

"Separatism," says a leader of a party committed to breaking Brazil apart, "is the only way for Brazil to shake off its backwardness." It may also be a path to civil conflict.

Across the world, then, we are hearing a premonitory growl from the angry affluent in an environment of clashing civilizations. The rich want out.

Many are thinking, if not saying aloud, "We can buy our needs and sell our goods abroad. Why saddle ourselves with an army of mal-nourished illiterates when our factories and offices might actually need fewer and higher-skilled workers in the future as the Third Wave advances?"

Whether such cleavages explode into violence, and how they might affect the major powers, will depend in part on how they intersect with the attempt to split the global economy into protectionist blocs.

THE ASIAN CHALLENGE

In the middle of the twentieth century, America, with the only Second Wave economy not shattered by World War II, had a virtual monopoly in many export items, from automobiles to household appliances, machinery, and other manufactured goods.

As Japan and Europe, with U.S. help, recovered from the war, they became competitive in a few lines of goods. But only in the seventies, when it began systematically introducing Third Wave production methods and transferring many Second Wave functions to less developed Asian economies, was Japan able to seriously invade U.S. and European markets with precision-manufactured goods of superb quality.

As Japan piled up enormous profits, it poured investment into many Southeast Asian countries, in turn stimulating their development. Soon these countries, too, became aggressive exporters, further stiffening global competition. Today, with coastal China coming on line, the battle for markets is becoming white hot. And it will become even more extreme as these countries, too, replace more and more of their Second Wave cheap labor factories with sophisticated Third Wave plants.

Faced by this powerful tide of competition, corporate forces in the

United States, with trade union backing, have orchestrated a massive propaganda campaign calling on Uncle Sam to protect or subsidize their domestic production. A parallel, even more intense campaign against Asian imports is under way in Europe.

THE FLAMING MATCH

Historians tell us that, as one country after another erected trade barriers in the 1930s, they savaged one another's economies, worsened unemployment, inflamed national passions, threw nations into political paroxysms, fueled Nazism and Stalinism, and lit the match that helped set the entire world aflame in the most destructive war in history. Yet today, even as economists and politicians invoke these memories and stress the danger of closed regional trading blocs, they prepare to construct them.

In no field is hypocrisy more shameless. The Japanese are past masters at limiting competition from abroad, pumping their own exports into every crevice of the world market, denying that they protect their markets, and simultaneously promising yet again to open them.

Conversely, the United States, for all its rhetoric about free trade and "level playing fields," imposes some 3,000 tariffs and quotas on everything from sweaters and sneakers to ice cream and orange juice. It negotiates free trade agreements with Canada and Mexico, in the process creating a zone that could someday be snapped shut against Asian exports and capital. And it engages in "currency protectionism" by promoting a low dollar, thus raising the cost of imports to the short-term advantage of domestic manufacturers. Europe, in turn, while haranguing against Japan, subsidizes its farmers, its aerospace and electronics industries, and engages in other spurious trade practices. Meanwhile, certain Southeast Asian nations mutter about creating their own bloc.

Economic arguments are increasingly buttressed by mutual bashing in the press, racist attacks, yellow-peril rhetoric, and other forms of hate-mongering with the potential for stoking violence. If massive markets are not quickly opened for previously nonexistent products like advanced environmental technologies, capitulation to protectionism, even under the guise of "managed trade" and other euphemistic formulas, could drive various nations to desperation and trigger disastrous confrontations in a world bristling with weapons as never before.

Dividing the Pacific into trade blocs, drawing what is in effect an ethno-racial line down its middle, could hack the most dangerous cleavage of all — racial, religious, and economic — into a global system already in danger of multiple fracture.

BACK FROM THE DEAD

All these tensions widen other global cleavages. The rise of religious fanaticism (as distinct from mere fundamentalism) promotes paranoia and loathing around the world. A minority of Islamic extremists conjure fantasies of a New Crusade, with the entire Muslim world united in a jihad, or Holy War, against Judeo-Christianity. On the other side, fascists in Western Europe pose as the last defenders of Christianity against a murderous Islam.

From Russia, where fascists wrap themselves in the flag of Orthodox Christianity, to India, where Hindu pogroms are carried out against Muslims, to the Middle East, where Iran promotes terror in the name of Islam, the world looks with wonderment at the multiplying millions who seem eager to hurl themselves back into the twelfth century.

This sudden, seemingly inexplicable resurgence of religion in general and fundamentalism in particular becomes comprehensible when seen in the context of clashing civilizations. When the Second Wave began spreading industrial civilization across Western Europe, the church, typically a great landowner, joined with First Wave agrarian elites against the rising commercial-industrial classes and their intellectual and cultural allies. The latter, in turn, attacked religion as a reactionary, anti-scientific, anti-democratic force, and made secularism the virtual hallmark of industrial civilization.

This great cultural war, which raged on for over two centuries, eventually resulted in the triumph of modernism — the culture of industrialism. With it came secular schools, secular institutions, and a generalized retreat of religion in the industrial countries. "Is God Dead?" asked *Time* magazine on its cover in April 1966.

Today, however, with Third Wave economies on the march, and Second Wave civilization in terminal crisis, secularism is caught in a pincer attack. On one side it is reviled by religious extremists who never gave up their hatred of modernity and wish to reinstate preindustrial fundamentalisms. On the other, it is attacked by the fast-multiplying "New Age" spiritual movements and religions, many of them essentially pagan, but religious, nevertheless.

Both at home and in the world at large, therefore, Second Wave secularism is thus no longer automatically regarded as the advanced, progressive philosophy of the future.

On a world scale, the lurch back to religion reflects a desperate search for something to replace fallen Second Wave faiths — whether Marxism or nationalism, or for that matter Scientism. In the First Wave world it is fed by memories of Second Wave exploitation. Thus it is the aftertaste of colonialism that makes First Wave Islamic populations so bitter against the West. It is the failure of socialism that propels Yugoslavs and Russians toward chauvinistic-cum-religious delirium. It is alienation and fear of immigrants that drives many Western Europeans into a fury of racism that camouflages itself as a defense of Christianity. It is corruption and the failures of Second Wave democratic forms that could well send some of the ex-Soviet republics tracking back either to Orthodox authoritarianism or Muslim fanaticism.

But religious passions, whether genuine or a mask for other sentiments, can be stoked by political demagogs and all too easily converted into a fever for violence. The ethno-religious nightmare in the Balkans merely foreshadows what might easily happen elsewhere.

THE UNCONTAINABLE REVOLUTION

These multiplying, fast-widening cleavages represent large-scale threats to peace in the decades ahead. They derive from the master conflict of our era — sparked by the rise of a revolutionary new civilization that cannot be contained within the bisected structure of world power that sprang up after the industrial revolution.

What we will see in the decades to come is a gradual trisection of the world system into First Wave, Second Wave, and Third Wave states, each with its own vital interests, its own feuding elites, its own crises and agendas. This is the grand historical context in which we observe the civilianization of war, the proliferation of nuclear, chemical, and biological weapons, and of missiles, and the rise of a completely unprecedented Third Wave war-form.

We are racing into a strange and novel period of future-history. Those who wish to prevent or limit war must take these new facts into account, see the hidden connections among them, and recognize the waves of change transforming our world.

In the period of extreme turbulence and danger to come, survival will depend on our doing something no one has done for at least two centuries. Just as we have invented a new war-form, we will have to invent a new "peace-form."

And that is what the remaining pages of this work are about.

PART SIX

PEACE

23

ABOUT PEACE-FORMS

ONE of the most famous combat stories in all of Western culture is the biblical tale of David, the Israelite, and Goliath, the Philistine. In it, the underdog David slays his giant antagonist with the help of a high-tech weapon — the slingshot.

Their duel exemplifies one of the lifesaving methods introduced by primitive people to minimize the effects of violence. Instead of entire tribes or clans tearing one another to shreds, many primitive groups settled their disputes by staging single combat — the choice of one champion to represent each side.

In Homeric legend, Menelaus, for the Greeks, and Paris, for the Trojans, fight a similarly decisive duel. Anthropologists have found evidence of single combat among Tlingit tribesmen in southern Alaska, Maoris in New Zealand, and other communities from Brazil to Australia.

Another lifesaving social innovation among certain primitive tribes was exemption — for example, sparing women and children, or neutrals or messengers sent by the enemy. A third idea exempted not people but certain places (in the New Hebrides, we learn, warring tribes set aside an inviolable "path of peace"). A fourth set aside certain times when fighting must stop — time out, for example, for religious ceremonies to take place.

As First Wave civilization arose, it brought into being a characteristic peace-form to match its war-form — a new set of tools, that is, for preventing war or mitigating violence.

For instance, the First Wave revolution that raised war above the level of tribal skirmishes also changed the fate of captives. Until then, live prisoners were of no use to the victorious tribe, except perhaps as replacements for fallen warriors or women needed for reproduction. Once agriculture made it possible to create food surpluses, however, and prisoners could generate more food than required to keep themselves fed, it became more profitable to enslave than to eat or kill them. Horrible as slavery was, it was one of many First Wave innovations that had the effect of reducing the battlefield body count. It was part of the peace-form of First Wave civilization.

The same thing happened when the industrial revolution arrived: Second Wave civilization, too, created its own war-form — and a peace-form to match.

For example, industrialism, when it first arose in Western Europe, placed a heavy emphasis on contractual relationships. Contracts became a part of everyday business life. Political systems were typically justified in terms of a "social contract" between the leaders and the led. It was a natural step for Second Wave nations to sign contracts with one another. Treaties and agreements thus became key elements in the Second Wave peace-form. Some set ethical limits on the behavior of the soldier.

While "humanitarian ideas have existed for thousands of years . . . ," states a report of the Department of Peace and Conflict Research at Uppsala University, in Sweden, "only in the 17th and 18th centuries did governments in Europe issue 'Articles of War' establishing certain normative standards for the treatment of belligerents."

These codes laid the basis for a patchwork of treaties, customs, and judicial decisions. In 1864 nations agreed to regard battlefield doctors and nurses as neutrals, and to care for wounded and sick troops irrespective of nationality. In 1868 nations ruled certain explosive bullets out of bounds.

In 1899 the First Peace Conference in the Hague discussed (but did not accept) a moratorium on arms. It did, however, impose restraints on weapons and methods of war, such as the use of projectiles dropped from balloons, and it set up a court for the arbitration of disputes among nations.

Since then the world has negotiated treaties, covenants, and other agreements to ban or restrict chemical and bacteriological weapons, to further humanize the treatment of POWs, to prevent genocide and limit nuclear weapons. But the industrial imprint on "peace-work" went much deeper than contractual arrangements.

The modernizers who built Second Wave societies created national markets and gave birth to what we now think of as the nation-state. War grew from conflicts between city-states or royal families to violence organized by full-fledged nations — with governments in control of integrated, nation-sized economies.

Modernizers rationalized tax collection (providing national governments with the funding for bigger wars), linked their populations together with national transportation and communications systems, and filled the heads of the people with nationalist propaganda pumped out by their intellectual collaborators and the national media.

They also created completely new institutions to keep the peace. In doing so, not surprisingly, they focused their efforts on nations.

The League of Nations after World War I and the United Nations after World War II differed in many respects. But both were built around nations. Both the League and the UN recognized national sovereignty, the inviolability of each national border, and the right of independent nations (and only nations) to be fully represented in them.

The very concept of "national security," in whose name the vast military buildup of the past half century took place, reflects an emphasis on peace and security at the level of nations, as distinct from peace within nations or peace at the level of religions, ethnic groups, or civilizations.

The League of Nations, hailed in its time as the hope of humanity, shriveled to insignificance in the 1930s and did little to prevent World War II. The United Nations, paralyzed by the Cold War for most of its existence, has now begun to come out of its coma at precisely the moment when its fundamental unit — the nation-state — is becoming less, not more, important in the global order. And, of course, the kind of war these institutions were primarily designed to prevent were Second Wave wars of mass destruction.

Thus, Second Wave civilization, exactly like First Wave civilization before it, invented a peace-form in step with its characteristic war-form.

Precisely as is the case with the war-form, the creation of a new peace-form doesn't do away with an older one. But a new war-form creates new threats to peace, thus calling into being, usually after a very long lag time, a new peace-form that corresponds to the new conditions and to the character of the corresponding civilization.

The crisis the world faces today is the absence of a Third Wave peace-form that corresponds to the new conditions in the world system and to the realities of the Third Wave war-form.

24

THE NEXT PEACE-FORM

MAKING PEACE cannot depend on the prior solution of all the world's moral, social, and economic ills. Those who tell us that war is a result of poverty, injustice, corruption, overpopulation, and misery may be right, though the formula seems oversimple. But if these must be eliminated before peace is possible, then war prevention or limitation becomes a utopian exercise.

The problem is not how to promote peace in a perfect world but in the world that we actually have and the new one we are creating. In today's real world we have a new global system in the making and a brand-new way of making war, yet so far few corresponding innovations in the way we try to make peace.

In 1931, a British writer, A. C. F. Beales, opened his book *The History of Peace* with the observation that "every single idea current today about peace and war was being preached by organized bodies over a century ago." He was referring to the time when the first "peace societies" were formed in England in 1815. They sprang to life in precisely the period when the Second Wave war-form was being rapidly developed and extended by Napoleon, and over the years they, in turn, helped develop what became the Second Wave peace-form. But the most fundamental assumptions on which that peace-form was built are no longer tenable.

For example, the Second Wave idea that national governments are the only ones that can wield military force is now obsolete. We increasingly see military units that have broken free of central government control. Some, as in Russia, have reportedly come under the de facto control of local business interests. Others, as in the drug regions, may sell out to criminal syndicates. Still others work for ethnic or religious movements. And others operate independently of any external authority. Some, like the Bosnian Serbs, fall halfway in between. As the Third Wave spreads, we may see even more variations. But if the nation-state is losing its "monopoly of violence," who exactly are the new threats to peace? What kind of global order can accommodate de-monopolized violence?

Second Wave anti-war activists have spent whole generations campaigning against the military-industrial complex. But what happens when that converts, as we have seen, into a civilian-military complex? Does one mount a political campaign, picket signs and all, against the manufacturer of some perfectly innocent civilian product that just happens to have a military use?

Peace campaigners during the Second Wave period typically opposed arms exports. But it now turns out that Second and Third Wave arms are very different. Should arms designed for indiscriminate slaughter be lumped together with arms designed to minimize collateral casualties? If that distinction is ignored, might we not overlook important ways to reduce bloodshed in the years to come?

To oppose war itself is morally satisfying. But with a world fast dividing into First, Second, and Third Wave civilizations, three distinctly different forms of warfare need to be averted or limited, along with various combinations. Each may require a different set of responses from peacekeepers or peacemakers.

Then there is the United Nations, on which so many millions around the world pin their earnest hopes for peace. To assume, as many do, that peace would be served if the UN had its own, permanent, all-purpose military arm, rather than ad hoc modular forces custom-tailored for each mission, is to apply anachronistic Second Wave thinking. The variety of wars requires a variety of anti-war forces, not a single omnipurpose unit.

It is, unfortunately, equally naive to assume that the UN, given its present structure, could douse the flames of war if only it had adequate financial support. There are too many things the UN cannot do, and could not even if it had all the money it wants.

The very fact that the UN consists exclusively of nation-states is a straitjacket in today's world. The fact that the UN may work with private nonprofit agencies in disaster zones, for example, or that it extends "consultative" status to nongovernmental organizations masks the larger reality: these NGOs or nonstate actors are still regarded by the UN as a nuisance at best, a rival source of power at worst. In Bosnia, according to National Public Radio, UN forces refused to protect a humanitarian aid convoy organized jointly by Catholic and Muslim relief organizations. The Blue Helmets explained that their mandate did not extend to the protection of private-agency efforts. Yet in a world in which nonnational forces exercise increasing power, peace cannot be made or kept without them. If the UN is to work effectively in the Bosnias or Cambodias of the future, it will have to share power at the highest level with these nongovernmental organizations, not to mention global corporations and other entities. They will have to participate fully in the formulation of UN strategies for peace.

If the UN dinosaur cannot transform itself from a Second Wave bureaucratic organization to a more flexible, Third Wave organization that represents nonstate actors along with nations, competing centers of global power will be organized — "para-UNs" made up of these various excluded groupings.

DIPLO-DITHER

Second Wave assumptions and institutions helped paralyze the world when confronted with the recent violence in the Balkans, with all its atrocities, mass rapes, and Nazi-like "ethnic cleansing." That war is worth examining briefly here, since it is a possible model for others still to erupt.

What the world witnessed in the Balkans was, in part, a First Wave war, fought by ill-armed, ill-trained, hastily organized, and undisciplined irregulars. Some were supported by elements of the Second Wave military of the former Yugoslavia. The UN wasn't about to fight. Europeans and Americans were unwilling to wage either a First or a Second Wave war, arguing that the Balkans were simply a quagmire.

But no attempt was made to exploit the Third Wave war-form, which, as we'll see in a moment, might have reduced the slaughter. We saw, instead, strategic myopia, moral hypocrisy, futile wrangling about the uses of air power, and endless diplo-dither.

Assuming the outside world really did wish to stop the horrors of that war (which is at least questionable), the issue was never whether or not air power could have helped snuff out the fighting. The real issue was not air, ground, or sea, but First, Second, or Third Wave. As we will see, there were, in fact, things that could have been done to minimize the tragedy without risking either ground troops or pilots.

We saw no imagination — no thinking outside the conventional Second Wave frame of reference. Even assuming ground troops were needed, many options remained unexplored. If, for political reasons, they could not come from either the UN or from Europe or America, were there no alternatives?

PEACE, INC.

Why not, when nations have already lost the monopoly of violence, consider creating volunteer mercenary forces organized by private corporations to fight wars on a contract-fee basis for the United Nations — the *condottieri* of yesterday armed with some of the weapons, including non-lethal weapons, of tomorrow?

Governments unwilling to send their own young men and women to die in combat against Serbian, Croat, or Bosnian irregulars, including rapists and genocidal thugs, might have had fewer reservations about allowing the UN to contract with a nonpolitical, professional fighting force made up of volunteers from many nations — a rapid-deployment unit for hire. Or one under contract to the UN alone.

Of course, to prevent such companies from becoming wild cards, strict international ground rules would have to be set — transnational boards of directors, public monitoring of their funds, perhaps special arrangements to lease them equipment for specific purposes, rather than allowing them to build up gigantic warstocks of their own. But if governments cannot directly do the job, the world may well turn to corporations that can.

By contrast, one might also imagine the creation, someday, of internationally chartered "Peace Corporations," each of which is assigned some region of the globe. Instead of being paid for waging war, its sole source of profit would come from war limitation in its region. Its "product" would be reduced casualty numbers as measured against some recent base-line period.

Special, internationally sanctioned rules could permit these companies wide military and moral latitude to conduct unorthodox peacekeeping operations — to do what it takes, ranging from legalized bribery to propaganda to limited military intervention, to the supply of peacemaking forces in the region. Private investors might be found to capitalize such firms if, say, the international community or regional groups agreed to pay them a fee for services plus bonanza profits in years when casualties decline. And if this doesn't work, perhaps there are other ways to seed the world with highly motivated peace-preserving institutions. Why not make peace pay off?

Such ideas sound zany, and maybe they are. But, good or bad, they lie outside the common frame of reference and they are used here only to illustrate that once we think outside the conventional Second Wave framework, we may find imaginative alternatives to paralysis.

OPEN SKIES AND OPEN MINDS

Peace can sometimes be promoted by economic measures or imposed by force. But these are not the only available tools. Peace at the dawn of the twenty-first century requires the surgical application of a less tangible but frequently more potent weapon: knowledge.

Indeed, any thinking about peace that ignores the central economic resource of the Third Wave civilization — which is also the key to its military power — is by definition inadequate. After all, if at least some wars can now be won with information superiority, can anti-wars be won that way, too?

What is glaringly absent today, even as armies begin thinking strategically about the use of knowledge, are coherent knowledge strategies for peace.

Rudimentary elements of such a strategy have been in place for a long time, although not necessarily seen in relationship to one another. For example, the concept of "transparency."

This idea — that the open availability of military information might reduce suspicion and give all sides ample warning of threatening developments — lay behind the "Open Skies" proposal first made by President Dwight Eisenhower to Soviet premier Khrushchev at a summit meeting on July 21, 1955.

As a step toward reducing nuclear tensions and the danger of surprise attack, he proposed that the U.S. and U.S.S.R. "give each other a complete blueprint of our military establishments, from beginning to end, from one end of our countries to the other" and for each country to provide the other with facilities for aerial reconnaissance "where you can make all the pictures you choose and take them to your country to study."

The Soviets quickly rejected the idea. Nevertheless, from then on — in the same decades that saw advanced economies grow more and more information-intensive — we saw growing acceptance by many nations of surveillance, mutual monitoring, and data gathering, including the right of one country to make "intrusive" on-site inspections of another to verify compliance with arms-control agreements. For example, the 1971 Seabed Treaty permits either the UN or a signatory nation to demand verification. In 1986, thirty-five nations at the Stockholm disarmament conference agreed to open themselves to on-site, short-notice inspections, without right of refusal. Of course, the Iraq case demonstrates the weaknesses and resistance still facing outside inspectors. But the principle that data, information, and knowledge are needed to support peace — and that includes the right of access — is now embedded in international practice.

In 1989, President Bush resurrected Eisenhower's proposal. By now sophisticated satellites and sensors in the heavens could supplement aerial reconnaissance. The West thus offered a sweeping version of Open Skies plus on-site inspection of military facilities to cover not only the United States but Canada and Europe as well. The Russians were now ready to negotiate, they said, and they agreed to allow the use of synthetic aperture radar, which can "see" through any weather and operate at night as well. But they wanted to limit the detail that space-based sensors could define. While the West wanted to be able to spot items ten feet or larger, the Russians wanted to set the limit at forty feet.

But this entire negotiation is myopic. The sky, as we saw, is likely, in time, to be populated by many more surveillance satellites, including commercial ones, capable of seeing even smaller items right down to individual mortar tubes and hand-held weapons. The location of every Serbian, Croatian, or Bosnian gunner in the future will be identifiable. Bad weather and rough terrain will be less of an obstacle. Skies are going to be open whether governments want them to be or not. And not just skies. The subseas and the earth itself are going to be more transparent.

Instead of bewailing the cost of space-based surveillance technologies and sea and ground sensors, we need to see these as social expenditures vital to the preservation of peace. What are needed are agreements for widespread sharing of both the information they provide and of their cost. And where viable commercial markets are insufficient to spur their development, imaginative transnational forms, perhaps a mixture of public and private, can be created to accelerate development.

The exchange of data, information, and knowledge in a world increasingly marked by regional arms races is clearly a Third Wave tool for peace.

TRACKING TECHNOLOGY

Not all arms races lead to war — as the biggest one in history, that between the United States and the Soviet Union — proves. Intention, rather than mere capability, matters. But runaway arms sales, erratic buildups, sudden infusions of weapons into a tense region, and surprise shifts in military balances all raise unpredictability and hence the risks of violence. In light of this, the United Nations has proposed creating an "arms register," which would officially track arms exports and imports by participating governments. Some American arms-control advocates have suggested that the United States cut off aid to countries that refuse to report arms transfers to the UN.

The register idea has many holes in it. The most dangerous transfers are the ones least likely to be reported, and the idea once more assumes that governments are the only players that count. Nonetheless, the proposal indicates a further recognition of the importance of organized information to the maintenance of peace.

More, not less, information is also needed to slow further proliferation of weapons of mass destruction. Especially with the shift from single-purpose to dual- (or multiple-) use technologies, it is not just weapons that need to be tracked but the spread of technologies — including old ones.

In trying to determine whether Iraq was building nuclear weapons, the IAEA and otherwise intelligent nuclear experts were deceived not only by Saddam Hussein and by lack of tracking data but by an embarrassingly stupid assumption. They dismissed the idea that Iraq might use calutron technology to separate Uranium 235 from Uranium 238, since far more efficient ways of making weapon-grade material were now available. But Saddam pursued his drive along multiple tracks, one of which employed precisely the technology commonly regarded as obsolete in the high-tech world.

"It's astonishing," said Glenn T. Seaborg, a former chairman of the U.S. Atomic Energy Agency. "It's cataclysmic," said Leonard S. Spector, a nuclear expert at the Carnegie Endowment for International Peace. The most pungent comment came from J. Carson Mark, an ex-official at the Los Alamos lab where the world's first A-bombs were built. "Why spend all that money on intelligence," he wanted to know, "when it apparently and evidently learns nothing?"

If nothing else, the Iraqi experience should have proved that the best source of information about weapons proliferation often comes from inside. It was a defecting Iraqi who reportedly first tipped off the West to Saddam's use of calutrons.

If information increasingly lies at the heart of anti-war action, why not recognize its immense value? Why doesn't the Carnegie Foundation for International Peace, or some other foundation, or the United Nations, or, for that matter, the IAEA itself, announce to the entire world that it will give a bounty of $1 million to anyone who brings in credible evidence of nuclear smuggling or of weapons proliferation. The offer to make "instant millionaires" ought to bring in plenty of whistleblowers. A whistleblower prize might prove more effective, than some of the monitoring now supposedly protecting the world from atomic horror. If the IAEA isn't already buying such intelligence, why isn't it?

However, apart from trying to detect the spread of specified weapons, it will now be necessary to cast a far wider net and collect data on shipments of superseded as well as state-of-the-art materials and machines. This, in turn, poses difficult, if not insoluble, knowledge problems. For example, it may be more important to know what software a potential aggressor has than what hardware. What do we do then? Anti-warriors need to begin thinking about logics, languages, artificial intelligence, and even alternative epistemologies as they apply to peace.

Arms transactions in the future will also be haunted by a new concern — and compel us to rethink other stock attitudes as well. For example, who in the future will trust smart weapons acquired from others?

The day may come, if it hasn't already, when weapons could be sold with embedded components "smart" enough to limit (or prevent) their use under prespecified circumstances. American, French, or Russian arms manufacturers, or for that matter those of other advanced economies, could, for example, implant hidden self-destruct chips into exported planes, rocket launchers, tanks, or missiles — just in case the buyer ever becomes an enemy or resells the weapon to an enemy. Hidden instructions could eject a pilot from his fighter or make the plane explode. Future technologies based on global-positioning satellite data could conceivably program a weapons system to misfire or a navigational system to misfunction once it flies outside a set of geographical boundaries predetermined by the seller.

Are such speculations pure science fiction? Not according to a knowledgeable, high-level defense industry official. In fact, he told us, "we could have coded all the airplanes we sold. We could have put a tag or a recognizer in all the chips that operate airplanes we sold to the Middle East. . . . In the event of hostile action, we would be able to communicate to that chip and it will malfunction. This has to be happening in one form or another." This official was not the only one to tell us this.

Can the purchaser find the embedded component? Caveat emptor. "Very difficult," say the officials, "exceedingly difficult . . . close to impossible."

If true, it is an example of highly sophisticated knowledge warfare. But if arms manufacturers can partially lobotomize exports, might some computer "hackers" or "crackers" — supposedly in the interests of peace — access the manufacturing process and reprogram certain systems so they do not function in combat at all?

TOMORROW'S UNSOLVED MURDERS

There is also, as we saw earlier, the brain-drain problem, which is likely to grow. In the private sector a whole new body of law is springing up relating to intellectual property. General Motors sues a former executive for allegedly taking fourteen boxes of computer disks and documents with him to Volkswagen. IBM sues a former employee to prevent him from working for Seagate, a manufacturer of computer disk drives. These are attempts to regulate the flow of brain power among companies for purely commercial reasons.

The rivalry is just about money. At a far more serious level, we already see Western governments contributing funds to keep certain specialists employed in Russia so they don't emigrate to volatile countries carrying what is inside their skulls with them — nuclear know-how, for example.

But there is another, far more drastic form of knowledge control. In 1980 Yahva El Meshad was found dead in a room in the Hotel Meridien in Paris. In March 1990 another man, named Gerald Bull, was gunned down in Brussels. Both murders remain "unsolved" to this day.

It turns out, however, that El Meshad, an Egyptian, was a key figure in Saddam Hussein's drive to build a nuclear bomb, and Bull,

Canadian-born, was trying to build a "super-gun" for Saddam. As knowledge becomes even more economically and militarily valuable it is quite likely that other unsolved murders will occur around the world.

In an anarchic world, one can imagine countries, or even private organizations, putting a bounty on the head of certain technical specialists who lend their expertise to the construction of prohibited weaponry. Such assassinations could even be sanctioned by regional or global authority someday as being in the interests of peace — though it is far more likely they will occur "unofficially." One way or another, the management of knowledge flows will become an increasingly important issue for peace and peacemakers in tomorrow's anarchic turbulence.

TRADE-IN WEAPONS

The war- and peace-forms of tomorrow will pose excruciating moral questions and force hard decisions. For example, apart from trying to withhold certain kinds of technical knowledge from potential troublemakers, it may be sensible for the most technologically advanced nations to actually provide technical know-how to less-than-friendly states.

If some "pariah state" succeeds in developing weapons of mass destruction, the rest of the world faces a critical decision. Now that it has a weapon, do we want the proliferator government, however atrocious it may be, to keep the weapon under careful control, lest it fall into unauthorized hands? If so, should we actually make sophisticated control technologies like "Permissive Action Links" available to it? Or is it better to keep a "bad" government technologically ignorant even if that risks loss of control over its weapons of mass destruction? Again, we find the control of knowledge at the heart of the peace preservation process.

Furthermore, since knowledge-intensive Third Wave weaponry is more precise and, in theory, can kill and wound fewer soldiers and civilians than Second Wave weapons of indiscriminate destruction, would the world be better off if high-tech nations sold Third Wave weapons to less militarily advanced armies — taking back their Second Wave weapons as a trade-in and destroying them under international supervision? How about trade-ins for non-lethal weaponry?

Such ideas only hint at the bizarre issues that will face armies and peace advocates alike tomorrow.

When we speak of a knowledge strategy for peace, what role must training play? Should specialized international training centers be set up for soldiers assigned to the UN, or for other peacekeeping or disaster-relief functions? What about the application of sophisticated computer simulation for training in mediation, disaster relief, famine emergency, and cross-cultural conflict resolution?

Above all, what about the kinds of modeling, analysis, and data collection that will help shift the entire focus of anti-war action from present to future — anticipatory thinking, rather than crash efforts after first blood is drawn. This requires insight not merely into military balances, troop movements, and the like, but information about the political factions and structural pressures, the payoffs and constraints that drive decision making in each state.

Finally, and this takes us back to the Balkans, no knowledge strategy for peace can ignore one of the most important sources of information, misinformation, and disinformation, the media.

HOW TO START (AND NOT STOP) A WAR

European and American governments gave long lists of reasons why they would not risk either ground troops or pilots in defense of the suffering people of the Balkans, Bosnian, Croatian, and Serbian alike. But no government has yet explained why it failed to take completely safe, inexpensive measures to suffocate, or at least limit, that war.

Rather than some incomprehensible eruption of thousand-year-old hatreds among people who had lived together and intermarried in peace for generations, the war was deliberately ignited.

As Communist bosses in various parts of Yugoslavia became discredited in the post–Cold War era, they sought to hold on to power by switching from Marxist ideology to religio-tribalism. Irresponsible intellectuals, sucking up to power, provided them with theories of ethnic or religious superiority and plenty of hyper-emotional rhetorical ammunition. The media provided the artillery.

In the words of Milos Vasic, an editor of *Vreme,* the only independent magazine in Belgrade, the explosion of violence was "an artificial war, really, produced by television. All it took was a few years of fierce, reckless, chauvinist, intolerant, expansionist, warmongering propaganda to create enough hate to start the fighting."

To understand what happened, he told Americans during a wartime visit, "imagine a United States with every little TV station everywhere taking the same editorial line — a line dictated by David Duke.

You, too, would have war in five years." Albanian journalist Violeta Orosi agrees, saying that "the disintegration of Yugoslavia began as a media war."

In all the regions fanatics controlled the main media, censoring, destroying, or deliberately marginalizing moderates. Despite this, pro-peace groups and tiny newspapers and magazines struggled desperately to douse the flames of hatred. Vesna Pesic, director of the Center for Anti-War Action in Belgrade, pleaded for the outside world to acknowledge the existence of "those who do not support the policies of national hatred and war." There were peace marches in Belgrade. Even in Banja Luka, a Bosnian Serb stronghold, in the very midst of the fighting, a group of Bosnians, Serbs, and Croats formed themselves into an organization called Civic Forum to fight against ethnic and religious hatred.

Yet not one of the Western powers — the United States, France, Germany, England — let alone the rest of the world — gave financial or political aid to domestic opponents of a war whose bloodshed these same governments denounced daily. Nor did they or the United Nations devise anything approximating a media strategy for countering hate propaganda to moderate the violence.

Navy ships were stationed offshore to monitor an arms embargo. But with transmitters on the decks of navy ships or from the soil of nearby Italy or Greece, the UN itself could easily have provided a media voice for the silenced moderates in each region, injecting a stream of sanity to these former Yugoslav republics. Along with an embargo on arms, how about an embargo on hate propaganda? The UN or the great powers could have jammed local programming, if they chose. They could have also controlled all telecommunication and postal lines running into and out of the warring states. But none of this happened.

If U.S. psychological warfare experts in the Gulf could drop 29 million leaflets on the Iraqis, could a few thousand tiny, cheap radios, tuned to a "Peace Frequency" be dropped over the war zone so that combatants could hear something other than their own side's lies?

In the United States, Grace Aaron, cochair of the Los Angeles–based Peace Action, begged for the United States Information Agency to "start offshore news broadcasting to enable citizens of all former Yugoslav republics to hear balanced, accurate news reports on the war," not just in the combat zones but in Belgrade and Zagreb as well.

Others urged Radio Free Europe or Radio Liberty to take on this task. Where was the BBC? Or CNN? Or from peace-loving Japan, where was NHK? Simple translations of their regular broadcasts could have strengthened those who wanted to end the fighting.

It took two years after the outbreak of the war for the United States finally to announce that it would launch a Radio Free Serbia — but only on short wave, it being lamely explained that medium-wave radio would need bigger transmitters near the target area. In 1920 the Marconi Company in England broadcast a concert by Dame Nellie Melba that was heard as far away as Greece, but in 1993 it was somehow impossible to reach Zagreb or Belgrade from, say, Italy or from the oceans nearby. There were, by this time, fully 500,000 satellite dishes in Serbia and Montenegro and another 40,000 in Croatia, but no international agency took advantage of them.

In the digital age, as we speed toward global, interactive multimedia, and giant media conglomerates race to tap future communications technologies, peace propaganda is still in the age of short-wave radio.

Clearly what is needed, not just by the United States but by the UN itself, if the UN is going to continue the pretense of peacekeeping, is a rapid reaction contingency broadcasting force that can go anywhere, set up, and beam news to those cut off from it — and not just on radio, but television as well.

According to Aaron, who has produced five cable access TV programs about war and peace for airing in the United States, Balkan groups are "unbelieveably sophisticated about propaganda." She has been given propaganda videotapes by all three sides in the war, some of the tapes clearly doctored. Some were from Serbian television programs picked up by satellite in the United States and taped for distribution by American pro-Serb activists.

Despite persecution by fanatics and governments in each of the warring regions, journalists, TV commentators, camera crews, and others struggled to speak out. Says Aaron, "The peace groups and the peace media could at least have been given some equipment — laptop computers, Sony Hi 8 cameras, video recorders, laser printers, modems, software, and subscriptions to outside world information services."

The point she makes is broader than the Balkans. "We're going to see an epidemic of regional conflicts. It will bankrupt the high-tech nations if they try to put all these down with military force. Why not use 'smart weapons' for peace?"

Why not, for example, a television miniseries that, instead of drug dealers, pimps, gang members, and corrupt cops, makes a hero out of a UN Blue Helmet — or out of the heroic individuals who stand against ethnic cleansing at the risk of their own lives?

Knowledge weapons alone, even including the use of the media, may never suffice to prevent war or to limit its spread. But the failure to develop systematic strategies for their use is inexcusable. Transparency, surveillance, weapons monitoring, the use of information technology, intelligence, interdiction of communication services, propaganda, the transition from mass lethality to low-lethal or non-lethal weapons, training, and education are all elements of a peace-form for the future.

Although they often approach issues from diametrically opposed positions, there are times when the interests of armies and peace movements actually coincide. If there were moral and strategic reasons why the United States would have preferred stability to war in the Balkans, the military, in carrying out a knowledge strategy in pursuit of that objective, might have worked with American peace activists to support their beleaguered counterparts in the war zone. Peace organizers might have called on the military for ships on which to base broadcast transmitters or planes to help deliver communications equipment to Balkan moderates.

Indeed, there is a deeper level on which peace and peacekeeping depends on knowledge. In a paper prepared for a conference of U.S. military and intelligence experts, Dr. Elin Whitney-Smith, a director of Micro Information Systems, Inc., has argued, as we have in our own work over the years, that wide access to information and communication is a precondition for economic development. Since poverty is no friend of peace, she proposed using "our military and the power of the digital revolution to get as much information and information technology out to the rest of the world [as possible] so that people in underdeveloped nations can become part of the global community. . . .

"In the interests of national security," she continued, "we need to use this knowledge to bring prosperity to the rest of the world before all its people become immigrants, refugees, or pensioners of the West."

Her words no doubt sounded utopian to some ears. But it will take all the Third Wave ideas we can get, along with the efforts of peace advocates and soldiers alike, for us to survive the upheavals that lie ahead as the global system trisects.

The old world order, built over the industrial centuries, already lies in fragments. We have argued throughout that the rise of a new wealth creation system and a new war-form demands a new peace-form. But unless that peace-form accurately reflects twenty-first-century realities, it could prove to be not only irrelevant but dangerous.

To design a peace-form for the future, however, we need a preliminary map of the global system of the twenty-first century. That map will be traced in the few remaining pages of this book.

25

THE TWENTY-FIRST-CENTURY GLOBAL SYSTEM

F EW WORDS are more loosely thrown about today than the term "global." Ecology is said to be a "global" problem. The media are said to be creating a "global" village. Companies proudly announce that they are "globalizing." Economists speak of "global" growth or recession. And the politician, UN official, or media pundit doesn't exist who isn't prepared to lecture us about the "global system."

There *is*, of course, a global system. But it is not what most people imagine it to be.

Efforts to prevent, limit, end, or settle wars, whether by armies or peace activists or anyone else, require some understanding of the system within which the war is taking place. If our map of the system is obsolete, picturing it as it was yesterday, rather than as it is fast becoming, even the best strategies for peace can trigger the opposite. Twenty-first-century strategic thinking, therefore, must start with a map of the global system of tomorrow.

BLAMING THE END OF THE COLD WAR

Most attempts to map the system begin with the end of the Cold War, as though that were the main force changing it. The end of the Cold War is still having an impact on the global system. But it is the thesis of this book that the changes arising from the breakup of the Soviet Union are secondary, and that, in fact, the global system would be caught up in revolutionary upheaval today even if the Berlin Wall had not fallen and the Soviet Union still existed. Blaming all of today's upheavals on the end of the Cold War is a substitute for thought.

We are witnessing, instead, the sudden eruption of a new civilization on the planet, carrying with it a knowledge-intensive way of creating wealth that is trisecting and transforming the entire global system today. Everything in that system is now mutating, from its basic components . . . to the way they interrelate . . . to the speed of their interactions . . . to the interests over which countries contend . . . to the kinds of wars that may result and which need to be prevented.

THE RISE OF THE SOFT-EDGED STATE

Start with the components. For the past three centuries the basic unit of the world system has been the nation-state. But this building block of the global system is itself changing.

The startling fact is that of all the present members of the United Nations roughly a third are now threatened by significant rebel movements, dissidents, or governments-in-exile. From Myanmar, with its fleeing masses of Muslims and its armed Karen rebels, all the way to Mali, where the Tuareg tribe is demanding independence, from Azerbaijan to Zaire, existing states face prenational tribalism — even though the slogans may refer to nationhood.

In testifying before the U.S. Senate Foreign Relations Committee prior to taking office, Secretary of State Warren Christopher — surely no scaremonger — warned that "if we don't find some way that the different ethnic groups can live together in a country . . . we'll have 5,000 countries rather than the hundred-plus we now have."

In Singapore we spoke with George Yao, the Cambridge- and Harvard-educated deputy prime minister. A thirty-seven-year-old brigadier general with a laser-like intellect, Yao imagines a future China composed of hundreds of Singapore-like city-states.

Many of today's states are going to splinter or transform, and the resultant units may not be integrated nations at all, in the modern sense, but a variety of other entities from tribal federations to Third Wave city-states. The United Nations may find itself, in part, a club of ex-nations or *faux* nations — other kinds of political units merely dressed in the trappings of the nation.

But that is not the only change looming on the horizon. In the high-tech world, the economic basis of the nation is sliding out from under it. There, as previously noted, national markets are becoming less important than local, regional, and global markets. On the production side, it becomes nearly impossible to tell what country a particular car or computer comes from, since its parts and software come from many different sources. The most dynamic sectors of the new economy are not national: they are either sub-, supra-, or transnational.

What's more, while poor, powerless, and "wannabe" groups demand "sovereignty," the most powerful and economically advanced states of all are losing theirs. Even the most powerful governments and their central banks can no longer control their own currency rates in a world awash in unregulated tidal waves of electronic money. They cannot even control their borders as they might have in the past. Even when they try to slam the door shut to imports or immigrants — both painfully hard to do — the high-tech states find themselves increasingly penetrated from outside by flows of money, terrorists, guns, drugs, culture, religion, pop music, ideology, information, and much else besides. In 1950, 25 million people traveled outside the borders of their own nation. By the late 1980s that number had soared to 325 million a year — plus an unknown and unknowable number of illegals. The old hard edges of the nation-state are eroding.

Thus the most basic components of the global system, as understood until now, are breaking down. There are more states in the system, and not all of them, despite their rhetoric, are nations.

Some, like the shakier former Soviet republics in the Caucasus, are essentially prenational "wannabes," First Wave societies torn apart by local warlords. Another tier consists of Second Wave nations. And an emerging Third Wave tier consists of a new kind of political entity — "soft-edged" postnational states. What is actually happening is the shift from a global system based on *nations* to a three-tier system based on *states*.

THE HIGH-TECH ARCHIPELAGO

Soon to be included in the newest, third tier of the system are regional "technopoles." In the words of Riccardo Petrella, director of science and technology forecasting for the European Community, "transnational business firms . . . are creating . . . networks, which bypass the nation-state framework. . . .

"By the middle of the next century, such nation-states as Germany, Italy, the United States, or Japan will no longer be the most relevant socioeconomic entities and the ultimate political configuration. Instead, areas like Orange County, California; Osaka, Japan; the Lyon region of France; or Germany's *Ruhrgebiete* will acquire predominant socioeconomic status. . . . The real decision-making powers of the future . . . will be transnational companies in alliance with city-regional governments." These units, he says, could form "a high-tech archipelago . . . amid a sea of impoverished humanity."

These regional units are assuming economic viability in the places where the Third Wave is most advanced. They are less viable in Second Wave economies still built around mass manufacturing for their national market. They reflect the more decentralized character of First Wave societies — only now on a high-technology basis.

CEOS, MONKS, AND MULLAHS

Two other obvious contenders for power in the global system are the great transnational corporations and religions, both increasing in reach and scope. Corporations like Unilever, whose 500 sub-companies operate in 75 countries, or like Exxon, 75 percent of whose revenues come from outside the United States, or, for that matter, IBM, Siemens, and British Petroleum, can no longer simply be regarded as "national" companies.

AT&T, one of the world's largest telecom firms, estimates that there are 2,000–3,000 giant companies in need of its global services. The United Nations describes 35,000 firms as transnational corporations. These companies have among them 150,000 affiliates. So extensive has this network become that an estimated one quarter of all world trade now consists of sales between subsidiaries of the same firm. This growing, collective organism, no longer strongly tethered to the nation-state, represents a crucial element in tomorrow's global system.

Similarly, the growing influence of global religions, from Islam to the Russian Orthodoxy to the fast-multiplying New Age sects, needs hardly be documented. All will be key players in the world system of the twenty-first century.

FROM GOLFERS TO METALWORKERS

In addition to states, regional "technopoles," corporations, and religions, another type of unit is also growing in importance: thousands of transnational associations and organizations now springing up like mushrooms after rain. Doctors, ceramicists, nuclear physicists, golfers, artists, unionized metalworkers, writers, industrial groups from fields as diverse as plastics and banking, health lobbies, trade unions, and environmental groups all now have larger-than-national interests and their own global organizations and agendas. These NGOs, or nongovernmental organizations, play an increasingly active role in the management of the world system and include, as a special class, a host of transnational political movements as well.

An obvious example is Greenpeace, the heavily funded environmental organization. But it is only one of a growing number of such global political actors. Many of them are highly sophisticated, armed with computers and faxes, and enjoy access to supercomputer networks, satellite transponders, and all the other means of advanced communication. When skinheads in Dresden, Germany, trashed an immigrant neighborhood, news of the event was blitzed out over ComLink, an electronic net connecting about fifty local computer networks in Germany and Austria. From there it went into Britain's GreenNet, which in turn is connected to "progressive" networks from North and South America to the former Soviet republics. A bombardment of faxes protesting the attack from all over the world deluged Dresden's newspapers.

But transborder electronic networks are not the monopoly of peace advocates who oppose violence. Networks connect up everyone from ecological extremists to biblical inerrantists, Zen fascists, criminal syndicates, and academic admirers of Peru's Sendero Luminoso terrorists, all forming part of a rapidly proliferating "transnational civil society" that may not always act with civility.

Here, too, the global system is trisecting. Transnational organizations are weak or even nonexistent in First Wave societies. They are more numerous in Second Wave societies. They breed at extremely high speed in Third Wave societies.

In sum, the old global system built around a few neatly defined nation-state "chips" is replaced by a twenty-first-century global computer — a three-level "motherboard," as it were, into which thousands and thousands of extremely varied chips are plugged.

HYPER-CONNECTIONS

The components of the world system are also wired together in new ways. Conventional wisdom today keeps telling us that the nations of the world are growing more interdependent. But that is, at best, a misleading oversimplification. It turns out that some countries are hypo-connected to the rest of the world while others are, if anything, hyper-connected.

First Wave states may be heavily dependent on one or a few other countries to buy their agricultural goods and raw materials. Zambia sells its copper, Cuba its sugar, Bolivia its tin. But their economies typically lack diversification. One-crop agriculture, concentration on one or a handful of resources, a stunted manufacturing sector, and underdeveloped services all reduce the need for linkages to the outside world. Such countries typically remain low on the interdependency or -connectivity scale.

Second Wave countries, because their economies and social structures are more complex, need more varied connections with the outside world. Yet even among industrial nations global interdependency is limited. As late as 1930, the United States, for example, was a partner to only thirty-four treaties or agreements with other countries. In 1968, even after its transition to a Third Wave economy had begun, the United States was still bound by only 282 such "contracts." Smokestack nations are, in general, therefore, moderately interdependent.

The Third Wave, by contrast, forces high-tech countries toward hyper-connectivity. Internally, as we know, these countries are going through a painful process of economic deconstruction and reconstruction. Giant corporations and government bureaucracies reorganize, break up, or decline in importance. New ones arise to take their place. Small units of all kinds multiply and form temporary alliances and consortia, crisscrossing the society with plug-in, plug-out modular organizations. Markets fracture into smaller and smaller segments as the mass society itself de-massifies.

This internal process, described in greater detail in an earlier chapter, has, in turn, an impact on the society's external relations. As it unfolds, companies, social and ethnic groups, agencies, and institutions develop a vast number of varied connections with the outside world. The more heterogeneous they become, the more they travel, export, import, communicate, and exchange information with the other parts of the world, and the more joint ventures, strategic alliances, consortia, and associations they form across borders. They move, in short, into the stage of hyper-connectivity.

This explains why, starting in the 1970s, the number of cross-agreements and treaties between the United States and other countries began to grow exponentially. Today the United States is party to slightly over 1,000 treaties and literally tens of thousands of agreements, each no doubt viewed as beneficial, but each also imposing constraints on its behavior.

We see, therefore, a complex new global system made up of regions, corporations, religions, nongovernmental organizations, and political movements, all contending, all with different interests, all reflecting different degrees of interactivity.

Hyper-connectivity produces an amazing, overlooked paradox. Japan, the United States, and Europe need the most linkages, the most highly interdependent relations with the outside world to sustain their advanced economies. We thus create a very strange world in which the most powerful countries are also the ones most tied down by external commitments. In this surprising sense, the most powerful are the least free. Small states, less dependent on outside ties, may have fewer resources, but can often deploy them more freely — which is why some micro-states can run rings around the United States.

GLOBAL "CLOCK-SPEEDS"

Furthermore, even as we plug more varied components into the global "motherboard" and link them up differently, we are also resetting its inner clock. Thus the new global system operates, as it were, at three sharply different "clock-speeds."

Nothing marks today's moment of history off from the earlier periods more strikingly than the acceleration of change. When we first made that point in *Future Shock* many years ago, the world had still to be convinced that events were, indeed, speeding up. Today few doubt it. The sense that events are moving faster is palpable.

This acceleration, partly driven by faster communication, means that hot-spots can materialize and war erupt into the global system almost overnight. Dramatic events demand response before governments have had time to digest their significance. Politicians are compelled to make more and more decisions about things they know less and less about at a faster and faster rate.

But like connectivity, acceleration is not the same throughout the entire global system. The general pace of life, including everything from the speed of business transactions to the rhythms of political change, the pace of technological innovation, and other variables, is slowest in agrarian societies, somewhat faster in industrial societies, and races at electronic speeds in the countries transitioning to Third Wave economies.

These differences produce markedly different views of the world. For example, it is hard for most Americans, whose daily life is among the fastest on earth and whose time horizons are truncated, to empathize with the feelings of warring Arabs and Israelis who defend their positions by citing 2,000-year-old claims. For Americans, history vanishes into itself very quickly, leaving only the immediate instant.

Such differences in time-consciousness even affect strategic thinking about war. Aware of American impatience, Saddam Hussein believed that the United States could not endure a long war. (He may have been right. But what he got was a short one.) Similarly, as we've seen, the Third Wave war-form itself not only emphasizes temporal over spatial factors but depends heavily on speed of communication and speed of movement.

Put differently, we are constructing not only a trilevel global system but one that operates in three different speed bands.

SURVIVAL NEEDS

This trisection also changes the things countries will live or die for in the future. All countries seek to protect their citizens. They need energy, food, capital, and access to sea and air transport. But beyond these and a few other elementals, their needs diverge.

For First Wave economies, land, energy, access to water for irrigation, cooking oil, food in desperate times, minimal literacy, and markets for cash crops or raw materials are generally the survival essentials. Lacking industry and exportable knowledge-based services, they see their natural resources, from rain forests to water supplies to fishing fields, as their chief salable assets.

States in the Second Wave tier, still reliant on cheap manual labor and mass manufacture, are nations with concentrated, integrated national economies. Because they are more urbanized, they need heavy food imports, either from their own countryside or from abroad. They need high inputs of energy per unit of production. They need bulk raw materials to keep their factories going — iron, steel, cement, timber, petrochemicals, and the like. They are the home of a small number of global corporations. They are major producers of pollution and other ecological negatives. Above all, they need export markets for their mass-manufactured goods.

Third Wave "postnations" form the newest tier of the global system. Unlike agrarian states, they have no great need for additional territory. Unlike industrial states, they have no need for vast natural resources of their own. (Lacking these, Second Wave Japan seized Korea, Manchuria, and other resource-rich regions. Third Wave Japan, by contrast, has grown immeasurably richer without either colonies or raw materials of its own.)

Third Wave "postnations," of course, still need energy and food, but what they also need now is knowledge convertible into wealth. They need access to, or control of, world data banks and telecommunications networks. They need markets for intelligence-intensive products and services, for financial services . . . management consulting . . . software . . . television programming . . . banking . . . reservation systems . . . credit information . . . insurance . . . pharmaceutical research . . . network management . . . information systems integration . . . economic intelligence . . . training systems . . . simulations . . . news services . . . and all the information and telecommunications technologies on which these depend. They need protection against piracy of intellectual products. And, as for ecology, they want the "unspoiled" First Wave countries to protect their jungles, skies, and greenery for the "global good" — sometimes even if it stifles economic development.

The diverging needs of First, Second, and Third Wave economies are reflected in radically different conceptions of "national interest" (a term that is itself increasingly anachronistic) and which could produce sharp tensions among countries in the years to come.

When we now plug all these changes together — differences in the types of units that make up the system; in their connectivity; in their speed; and in their survival requirements — we arrive at a transformation that reaches far beyond anything made necessary by the end of the Cold War. We arrive, in short, at a twenty-first-century global system, the arena in which the wars and anti-wars of tomorrow will be fought.

THE END OF EQUILIBRIUM (NOT HISTORY)

Second Wave theories about the global system tended to assume that it is equilibrial, that it has self-correcting elements in it, and that instabilities are exceptions to the rule. Wars, revolutions, and upheavals are unfortunate "perturbations" in an otherwise orderly system. Peace is the natural condition.

This view of the global order closely paralleled Second Wave scientific notions about order in the universe. Thus nations were like Newtonian billiard balls that bounced off one another. The entire theory of "balance of power" presupposed that if one nation grew too powerful, others would form a coalition to counteract it, thus returning it to its proper orbit and restoring equilibrium once again.

A related set of assumptions is still widely held in the affluent West. It includes the liberal idea that no one really wants war . . . that, deep down, adversaries are mirror images of ourselves . . . that governments are inherently aversive to risk . . . and that all differences can be negotiated peacefully if opponents will only keep talking to one another because, in the end, the global system is essentially rational.

Yet none of these assumptions apply today. At times some governments do, in fact, want war even in the absence of external threat. (The Argentinian generals who started the Falklands/Malvinas War in 1982 did so for purely internal political reasons in the absence of any external threat whatever.) Many leaders are not risk-aversive but thrive politically on high risk. For them, nothing succeeds like crisis.

More and more players on the world stage take on the characteristics of what Yehezkel Dror, a brilliant Israeli policy scientist, once called "crazy states." This is especially the case when the global system is caught up in revolution.

What many foreign policy pundits still fail to appreciate is that when systems are "far from equilibrium" they behave in bizarre ways that violate the usual rules. They become nonlinear — which means that small inputs can trigger gigantic effects. A tiny number of negative votes cast in tiny Denmark was enough to delay or derail the entire process of European integration.

A "small" war in a remote place can, through a series of often unpredictable events, snowball into a giant conflagration. Similarly, a big war can result in remarkably little change in the overall distribution of power. The Iran-Iraq war of 1980–88 caused over 600,000 casualties — yet ended in a standoff. There is a decreasing correlation between the size of an input and the size of the output.

The world system is taking on Prigoginian characteristics — that is, it looks more and more like the physical, chemical, and social systems described by Ilya Prigogine, the Nobel-prizewinning scientist who first identified what he called "dissipative-structures." In these, all parts of the system are in constant fluctuation. Parts of each system become extremely vulnerable to external influences — a change in oil prices, a sudden surge in religious fanaticism, a change in the balance of weapons, et cetera.

Positive-feedback loops multiply — meaning that once set in motion, certain processes take on a life of their own, and, far from being stabilized, introduce even larger instabilities into the system. Ethnic vendettas generate ethnic battles that generate ethnic wars larger than a given region can contain. A convergence of fluctuations, internal and external, can lead to total breakdown of the system — or to reorganization at a higher level.

Finally, at this critical moment the system is anything but rational. It is, in fact, more susceptible to chance than ever, meaning that its behavior is harder, perhaps even impossible to predict.

Welcome, then, to the global system of the twenty-first century — not the neat New World Order once touted by President Bush or the post–Cold War stability promised by other politicians. In it we can see the powerful process of trisection at work, reflecting the emergence, in our lifetime, of a new civilization with its own distinct survival needs, its own characteristic war-form, and soon, one hopes, a peace-form to match.

We live at a fantastic moment of human history. Hidden behind all the fashionable gloom today are several tremendously positive and humanizing changes on the planet. The spread of the Third Wave economy has galvanized all of the Asia Pacific region, introducing trade and strategic tensions, but at the same time opening the possibility of rapidly raising a billion human beings out of the pit of poverty. Massive increases in global population occurred between 1968 and 1990, but despite doomsday forecasts, per capita food supplies in the world have actually increased faster, according to the World Food Organization, and the number of chronically undernourished people has fallen by 16 percent.

Using Third Wave technologies that are less energy intensive and less polluting we can now begin to clean up the ecological havoc wrought by Second Wave industrial methods in the age of mass production. Work, until now brutalizing and mind-destroying for most of those lucky enough to hold a job, can be transformed into something fulfilling and mind-enhancing. The digital revolution that is helping to fuel the Third Wave has within it the potential for educating billions.

And despite the warnings throughout these pages about the danger of war, civil outbreaks, and even nuclear attacks, the good news is that, even though some 50,000–60,000 nuclear warheads have been produced since Hiroshima and Nagasaki, and even though there have been underground blasts and nuclear accidents, not one of those thousands of bombs has been detonated in anger. Some human survival instinct has repeatedly stayed the finger that might have pushed the button.

But to survive at the dawn of the twenty-first century will take more than instinct. For all of us, civilians and soldiers alike, it will take a profound understanding of the revolutionary new linkage between knowledge, wealth, and war. These pages will have served their purpose if they have illuminated that relationship. To make that happen, we have tried to sketch a new theory of war and anti-war. We will be gratified if we have brought one new insight to awareness or helped explode one obsolete idea that stands in the way of a more peaceful world.

We believe that the promise of the twenty-first century will swiftly evaporate if we continue using the intellectual weapons of yesterday. It will vanish even faster if we ever forget, even for a moment, those sobering words of Leon Trotsky's, quoted at the beginning of this book: "You may not be interested in war, but war is interested in you."

ACKNOWLEDGMENTS

Even More than most books, this one could not have been written without the help of numerous people. As outsiders to the military and the military culture, we were pleasantly surprised at the willingness of many officers, defense officials, academics, and others to speak with us about what we regard as the most dramatic upheaval in the nature of war and peace since the French Revolution. Everywhere we found deep questioning about what it will take to minimize violence in the decades ahead.

While it would be impossible to acknowledge all those we interviewed or discussed these matters with during the course of writing this book, several were unusually helpful. They include many high-ranking officials and officers, but we hope we will be forgiven for having taken the liberty of omitting their various titles and ranks, as these are changing faster than we can monitor.

Among those who took the time to assist us or share their ideas with us were Grace Aaron, Duane Andrews, John Arquilla, John Boyd, Carl Builder, Dick Cheney, Ray Cline, John Connally, Klaus Dannenberg, Michael Dewar, William Forster, Lewis Franklin, Pierre Gallois, Newt Gingrich, Dan Goldin, Daniel Goure, Jerome Granrud, Steve Hanser, Jerry Harrison, Ryan Henry, Zalmay Khalilizad, Tom King, Andy Marshall, Andy Messing, Janet and Chris Morris, Jim Pinkerton, Jonathan Pollock, Jonathan Regan, David Ronfeldt, Tim Rynne, Larry Seaquist, Stuart Slade, Donn Starry, Robert Steele, Bill Stofft, Paul Strassmann, Dean Wilkening, and Henry Yuen. As noted in the text, Patti Morelli, the widow of Don Morelli, was extremely kind as well.

Closer to home, we wish to thank our daughter, Karen Toffler, who, under difficult conditions, took responsibility for checking our research and preparing the Bibliography and Index. She worked indefatigably to meet the deadlines as they sped by. During the early months, Deborah Brown helped with the checking, until she had to leave to write her own book on a different topic. In the last-minute crush, Robert Basile bird-dogged items at the library and Valerie Vasquez aided in manuscript preparation. Of course, responsibility for any "stealth" errors that have crept into the text remains with the authors.

Throughout it all, Juan Gomez made sure that every piece of paper was where it belonged, that our cars and airplanes were there as needed, that interviews were correctly scheduled, that telephone calls and faxes from everywhere in the world were answered with intelligence, kindness, and good cheer. He helped in a thousand less noticeable, but equally important, ways as well.

The manuscript was greatly improved by Jim Silberman, our old friend and now our editor at Little, Brown. We received endless support from our agent, Perry Knowlton, and the people on his team at Curtis Brown Ltd., especially Grace Wherry, Dave Barbor, and Tim Knowlton.

NOTES

Bracketed [ ] numbers indicate items listed in the accompanying Bibliography. Thus, in the Notes [1] will stand for the first item in the Bibliography: *Bull's Eye,* by James Adams.

Certain sources have proved of special relevance, among them the consistently useful reporting in the weekly *Defense News* and the publications of the International Institute for Strategic Studies.

CHAPTER 1   UNEXPECTED ENCOUNTER

PAGE
   9   Biographical data about Brig. Gen. Don Morelli is based on material kindly supplied by his widow, Mrs. Patti Morelli, and by the U.S. Army Training and Doctrine Command, as well as on personal conversations with Morelli and interviews with officers who knew him.
   9   Third Wave: Our theory of waves of change is spelled out in [380] and [381].
10–11   Brain-force economy: [379], esp. chaps. 3–8.

CHAPTER 2   THE END OF ECSTASY

   13   Casualty figures: [2], p. 8; also "The Post Cold War and Its Implications for Military Expenditures in Developing Countries," by Robert McNamara, Paper dated January 25, 1991, esp. Appendix I.
   14   Three weeks of peace: "The 'Century of the Refugee,' A European Century?" by Hans Arnold, *Aussenpolitik,* No. III, 1991.
   15   Dismantling battleships: "Fulfilling the Treaty," by H. A. MacMullan, *Scientific American,* July 1922.
   16   Economic interdependence: [183], [317].
   16   Geo-economics: "America's Setting Sun," *New York Times,* September 23, 1991, and "U.S.-Japan Treaty Can Turn Things Around," *Los Angeles Times,* March 24, 1992, both by Edward Luttwak; also "The Primacy of Economics," by C. Fred Bergsten, *Foreign Policy,* Summer 1992 and [376], p. 23.

17   Zone of peace: "The Pentagon & Pax Americana," by Sol W. Sanders, *Global Affairs,* Summer 1992, p. 95.

CHAPTER 3   A CLASH OF CIVILIZATIONS

18   Civilization: As used by us throughout, this term refers to a way of life associated with a particular system for wealth production — agrarian, industrial, and now knowledge-based, or informational. In 1993 Samuel P. Huntington, director of the Olin Institute for Strategic Studies at Harvard, launched a discussion among American foreign policy specialists by announcing in the summer issue of *Foreign Affairs* and in the June 6th *New York Times* the decline of economic and ideological conflict in the world and the resurrection, in its place, of war between civilizations. In so doing, he challenged the geo-economic school, which sees trade conflict and global competition as the main source of future rivalries.

In his article, he identified "seven or eight major civilizations," which include "the Western, Confucian, Japanese, Islamic, Hindu, Slavic-Orthodox, Latin American and possibly African civilizations," adding that "the fault lines between civilizations will be the battle lines of the future." The dominant conflict, however, will be between "the West and the rest."

We, too, believe civilizations will clash in the future. But not along the lines he suggests. An even larger potential collision lies ahead — a "master conflict" within which his clash of civilizations could itself be subsumed. We might think of it as a collision of "super-civilizations."

While many civilizations and sub-civilizations have risen and fallen throughout history, there have been only two great "super-civilizations" into which all the others fit. One was the 10,000-year-old agrarian "super-civilization" that began the First Wave of change, and in time, had its Confucian, Hindu, Islamic, or Western variants. The other was the industrial "super-civilization" that swept a second wave of change across Western Europe and North America, and is still spreading to other parts of the world as well.

By the end of the nineteenth century pockets of industrialism had already appeared in Japan, Confucian China, and Slavic-Orthodox Russia as well. As the twentieth century unfolded, drives to industrialize (typically mis-identified as "Westernization") came to Muslim Turkey under Attaturk and to Iran under the Shah, to Catholic Brazil and Hindu India alike.

Each of these societies may have retained elements of their religion, culture, and ethnicity in their agrarian regions, but wherever industrial forces appeared they weakened these bonds. The spread of industrialism brought urbanization, much looser adherence to traditional religious and moral codes, and shattered many other cultural patterns as well. In short, the industrial super-civilization engulfed local civilizations wherever it spread.

Similarly, today's Third Wave civilization already has developed Western, Japanese, and Confucian versions. This is why we think the traditional definition of civilization on which Huntington relies is inadequate, and that the many clashes he foresees, if they occur, will occur within a much larger framework — a world increasingly divided into three distinct and potentially clashing super-civilizations.

Once this is grasped, we can simplify things in the pages ahead. We shall continue to use the word "civilization" to refer to First Wave agrarianism or Second Wave industrialism or to the emerging Third Wave society, and assume the adjectival "super" is understood.

19   On the industrial revolution: See [42], [59], [61], [82], [83], [113], [151], [152], [158], [189], [238], [277], [395], [398].
22   De-coupling: A further discussion will be found in [379], Chap. 30.

CHAPTER 4   THE REVOLUTIONARY PREMISE

29   Alexander: [115], p. 149.
30   Iphicrates: [115], p. 160.
30   Range estimates: see [99] and [44], pp. 35–36.
30   Pope Innocent II: [236], p. 68.
30   6,000 miles: [92], p. 7.
30   Laser: " 'Star Wars' Chemical Laser Is Unveiled," by Thomas H. Maugh, *Los Angeles Times,* June 23, 1991.
31   Kennedy: [72], p. 2.

CHAPTER 5   FIRST WAVE WAR

33   On tribal warfare: [86], p. 183.
33   War distinct from banditry: [38], p. 79.
34   Ancient China: Shang's largely overlooked manual is an astonishing document — rich with detailed observations and rules. Trenchantly logical and icily cruel, if Shang had been reincarnated in the twentieth century he might have been the deadliest adviser at Mao Tse-tung's side [110].
35   Greek fought Greek: [371], pp. 25–26; [144] Keegan Introduction and p. 35.
35   "The sovereign of a feudal country . . .": [397], p. 59.
36   Vassal obligations: [148], p. 64.
36   "blows, wounds, hard winters . . .": [95], p. 179.
36   Frederick the Great: [77], p. 17.

CHAPTER 6   SECOND WAVE WAR

38   After 1792: "Frederick the Great, Guibert, Bulow: From Dynastic to National War," by R. R. Palmer, in [278], p. 91.
39   Conscription in U.S. and Japan: [154], p. 432 and [193], p. 216; [145] pp. 22–23.
39   Whitney's muskets: [249], pp. 136–38.
40   Japanese army evolution: [145], p. 47.
40   U.S. World War II industrial base: [298], esp. pp. 880–81; [154], p. 787; see also "The Face of Victory," by Gerald Parshall, *U.S. News & World Report,* December 2, 1991.
41   Tokyo air raids: [176], p. 42.
41   Ludendorff and total war: "Ludendorrf: The German Concept of Total War," by Hans Speier, in [111], pp. 306–19.

CHAPTER 7   AIRLAND BATTLE

44   Starry profile: Interviews with Starry. Also: [71], pp. 244–45.
47   Yom Kippur War: Description of Golan Heights battle drawn from [173], [320], [150], [73], and interviews with Starry.
51–55 History of AirLand Battle: Interviews with Starry, supplemented by [316]; also "The Army Does an About-Face," by John M. Broder and Douglas Jehl, *Los Angeles Times,* April 20, 1991; "Joint Stars in Desert Storm," by Thomas S. Swalm, in [53], pp. 167–68; See also: [410].
55   1993 doctrinal revision: [411].

CHAPTER 8   THE WAY WE MAKE WEALTH ...

58   Knowledge-intensivity in economy: [379], esp. chaps. 3–8.
59   New products: "New Products Clog Groceries," by Eben Shapiro, *New York Times,* May 24, 1990.
61   IBM: "GM and IBM Face That Vision Thing," by James Flanigan, *Los Angeles Times,* October 25, 1992.
62   Nabisco: "Technology Helps Nabisco Foods Gain Order in a Turbulent Business," *Insights* (Computer Sciences Corporation), Spring 1991.
63   Vice President Gore: "The Information Infrastructure Project," Science, Technology, and Public Policy Program, John F. Kennedy School of Government, Harvard University, May 26–27, 1993; Statement of John H. Gibbons, Director, Office of Science and Technology Policy, the White House, about the "High Performance Computing and High Speed Networking Applications Act of 1993," April 27, 1993; "High-Speed Computer Networks Urged as Boon to Business, Schools," by Lee May, *Los Angeles Times,* November 21, 1991.

CHAPTER 9   THIRD WAVE WAR

65   Sample of exaggerated Gulf War casualty forecast: "War Toll Estimate: Up to 30,000 GIs in 20 Days," by Jack Anderson and Dale Van Atta, *Washington Post,* November 1, 1990.
65   Sample of technological pessimism: "Is Our High-Tech Military a Mirage?," by Harry G. Summers, *New York Times,* October 19, 1990.
66   Iraq jeep journey: Interview with Gallois.
67   Re 117s: [407], pp. 99, 116, 702–3.
69   Campen: [53], pp. ix–xi, 32–33.
70   Higher levels of command: "Rapid Proliferation and Distribution of Battlefield Information," by Timothy J. Gibson, in [53], p. 109.
70   J-Stars: "Joint Stars in Desert Storm," by Thomas S. Swalm, in [53], pp. 167–69.
70   Earliest targets: [407], p. 96.
71   Mernissi: [240], p. 43.
74   Question authority: "When the Anti-Military Generation Takes Office," by Steven D. Stark, *Los Angeles Times,* May 2, 1993.
74   Educated generals: "They Can Fight, Too," *Forbes,* March 18, 1991.
74   The human element: "Combat Enters the Hyperwar Era," by Lt. Col. Rosanne Bailey and Lt. Col. Thomas Kearney, *Defense News,* July 22, 1991.
75   "Not a mere ammunition mule": "Don't Call Today's Combat Soldier Low Skilled" (letter), by Col. W. C. Gregson, *New York Times,* February 19, 1991.

Raised skill requirements are matched by a need for new human relations skills as well — something the U.S. military is struggling with painfully. Day after day revelations of sexual harassment of women and mistreatment of homosexual soldiers within the armed forces show how deeply entrenched old "macho" attitudes remain in the military culture. In a rapidly diversifying Third Wave society, however, the military, like the new workforce, must learn to turn heterogeneity to advantage.

The American military has done a better job of reorganizing and changing the distribution of skills than many businesses, but it has so far done a worse job than many firms in challenging old values. As morale, adaptability, innovation, and technical knowledge become more important for survival, the advanced military will have to shed vestiges of patriarchy and intolerance based on race, religion, or sexual preference.

77   Opposite of micromanagement: [349], pp. 149–50.
77–78   Soviet "command from the rear": [346], p. 43.
78   Pagonis role: "General's Star Feat: Desert Armies Come, and Go," by Youssef M. Ibrahim, *New York Times,* November 8, 1991.
79   118 mobile stations: "Communications Support for the High Technology Battlefield," by Larry K. Wentz, in [53], p. 10.
79   700,000 telephone calls: "Desert Storm Communications," by Joseph S. Toma, in [53], p. 1.
80   Acceleration: "The Gospel According to Sun Tzu," by Joseph J. Romm, *Forbes,* December 9, 1991.

CHAPTER 10   A COLLISION OF WAR-FORMS

81   Clausewitz on the "big picture": [64], p. 584.
82   Machine guns: [113], pp. 86–87.
83   Satsuma rebellion: [145], pp. 30–32. See also [403].

CHAPTER 11   NICHE WARS

90   Keyworth quote: Proceedings of symposium of Open Source Solutions, Inc., Washington, D.C., December 1–3, 1992, Vol. I.
91   Special Operations in Gulf War: [407] pp. 114–15, 530, 532; also [330], p. 414.
91–92   Numbers and types of SO forces: [197], [302], [11].
92   Lobbyist for SO: Interviews with Messing.
93   Technology in first SO raid in Gulf War: [407], p. 115.
93   Iran hostage mission: [386], p. 77.
94   Special Ops Expo: Interviews with Bumback.
94   Old Colony meeting: Conference on Special Operations, Low Intensity Conflict and Drug Interdiction, November 7–8, 1991.
95   Simpson and Childress: at Old Colony conference.
96   Shachnow technology timeline: July 1992 presentation at John F. Kennedy Special Warfare Course and School, Fort Bragg, N.C.

CHAPTER 12   SPACE WARS

The First Information War, edited by Alan D. Campen [53], is an indispensable source of technical information about the Persian Gulf War, especially with respect to space.

98   First instance: [53], p. 135.
98   Anson and Cummings: "The First Space War" in [53], pp. 121–34.
98   Satellites and functions: "Military Space Program Faces a Reality Check," by Ralph Vartabedian, *Los Angeles Times,* October 30, 1992.
99   UN space agency: " 'Space Benefits' — A New Aspect of Global Politics," by Kai-Uwe Schrogl, *Aussenpolitik,* No. IV, 1991, pp. 373–82.
100   Missile launches: "SDI and Missile Proliferation," by John L. Piotrowski, *Global Affairs,* Spring 1991, p. 62.
100   North Korean missiles: "North Korea Alarms the Middle East," by Kenneth R. Timmerman, *Wall Street Journal Europe,* May 29–30, 1992; also "N. Korea Considers Scud Export Boost," by Terrence Kiernan, *Defense News,* April 26–May 2, 1993.

100   Missile Technology Control Regime: [283], p. 131.
101   Satellite proliferation: "Concern Raised as Emirates Seek Spy Satellite from
       U.S.," by William Broad, *New York Times,* November 17, 1992; "UAE
       Satellite Plan Rattles U.S.," by Vincent Kiernan and Andrew Lawler, *Defense News,* November 16–22, 1992.
102   Shift to "Ballistic Missile Defense": "The Rise and Fall of Strategic Defense"
       and "BMD Era Requires Vision, Difficult Choices," by Barbara Opall,
       Defense News, May 17–23, 1993; " 'Star Wars' Era Ends as Aspin Changes
       Focus," by Melissa Healy, *Los Angeles Times,* May 14, 1993.
102   Horner warning: "U.S. Space Warfare Chief Pleads for Orbiting Interceptor," by Barbara Opall, *Defense News,* May 10–16, 1993.
102   British plans: "Defence Ministry Considers Arming Against Third World
       Missile Risk," by Michael Evans, *The Times* (London), October 28, 1992.
102   French plans: "U.S., France Discuss Joint ATBM Program," by Giovanni de
       Briganti, *Defense News,* September 2, 1991.
102   Western European Union: "Europe Eyes Missile Defense," by Keith Payne,
       Defense News, May 24–30, 1993.
102   Need for anti-satellite weapon: "McPeak Presses for ASAT Option," by Neff
       Hudson and Andrew Lawler, *Defense News,* April 19–25, 1993.
103   Not-so-distant future: "After the Battle," by Eliot Cohen, *New Republic,*
       April 1, 1991.
103   Soviet ASATs announced: [46], pp. 76, 91.
103   ASAT tests: "A Response to the Union of Concerned Scientists," by Robert
       daCosta, *Defense Science,* August 1984.
106   Space "heartland": [72], pp. 1, 23, 47–49.

CHAPTER 13   ROBOT WARS

A primary source on military robotics is *War Without Men,* Vol. II, Future Warfare
Series. (Washington, D.C.: Pergamon-Brassey's, 1988), by Steven M. Shaker and Alan
R. Wise.
109   Casualty "standard": Interview with Harrison.
110   Driverless tank: Interview with Yuen. Also: "Lessons Learned from the Middle East War — Proposed Emphasis of Future Research," TRW Memo
       from Yuen, dated March 6, 1991.
110   A-Team: Interview with Harrison.
110   Meieran paper: "Roles of Mobile Robots in Kuwait and the Gulf War: What
       Could Have, Might Have, and Should Be Happening," Proceedings Manual, 18th Annual Technical Exhibit and Symposium, Association for Unmanned Vehicle Systems, Washington, D.C., August 13–15, 1991.
112   Japanese robot planes: "New Copter Able to Fly Pilotless," by Sumihiko
       Nonoichi, *Japan Economic Journal,* March 30, 1991.
113   Prowler material: [339], pp. 52–54.
114   Robo-terror: [339], p. 169.
114   TRW's Stone: "From Smart Bombs to Brilliant Missiles," by Evelyn Richards, *Washington Post National Weekly,* March 11–17, 1991.
115   Anti-robot sentiment: [339], pp. 170–71.
117   Artificial life: "A-Life Nightmare," by Steven Levy, *Whole Earth Review,*
       Fall 1992.

CHAPTER 14   DA VINCI DREAMS

118   Sensors and "smart" mines: Interview with Forster.

119 "Smart" armor: "DoD Probes Smart Tank Armor," by Vago Muradian, *Defense News*, March 1–7, 1993.
119 All-electric battlefield and exo-skeletal suit: Interviews with Harrison and Forster.
120 Micro-machines: "A Robot Ant Can Be Tool or Tiny Spy," by Edmund L. Andrews, *New York Times*, September 28, 1991.
120 Nano-technology: [104], [308], p. 362; see also letters by K. Eric Drexler, Susan G. Hadden, and Jorge Chapa in *Science*, January 17, 1992.
121 Chemical and biological weapons: See Chemical Disarmament and International Security, *Adelphi Papers* 267, International Institute for Strategic Studies; NBC Defense and Technology International, April 1986; "U.S. Studies of Biological Warfare Defense Could Have Offensive Results," *Discover*, June 1986; "Soviet Prods West on Exotic Weapons," *New York Times*, August 11, 1976; also [318] for a defense of chemical weaponry.
122 Racial weaponry: "Race Weapon Is Possible," *Defense News*, March 23, 1992.
123 Romans in Carthage: [138], p. 144.
124 30 years from now: Testimony of Alvin Toffler, U.S. Senate Committee on Foreign Relations, 94th Congress, First Session, Hearings May 7–June 4, 1975; excerpted in *International Associations*, 1975, p. 593.

CHAPTER 15   WAR WITHOUT BLOOD?

126 U.S. Global Strategy Council role in non-lethality: Interviews with Cline.
126 Non-lethality project: Interviews with Morrises.
126 For general description of USGSC project: "Nonlethality: Development of a National Policy and Employing Nonlethal Means in a New Strategic Era," a U.S. Global Strategy Council paper, n.d.
127 Definitions of non-lethality: Interviews with Morrises; USGSC project documents.
127 Opposition to perverted versions of non-lethality: Interviews with Morrises.
128 War can never be humane: "In Search of a Nonlethal Strategy," by Janet Morris, a USGSC non-lethality project paper, n.d.
128 Reactions to project: "Futurists See a Kinder and Gentler Pentagon," *San Francisco Examiner*, February 16, 1992.
128 Perry Smith comment: [349], p. 141.
128 Warden quoted: "Pentagon Forges Strategy on Non-Lethal Warfare," by Barbara Opall, *Defense News*, February 16, 1992.
130–131 Laser weapons: "Soviet Beam Weapons Are Near Tactical Maturity," by Lt. Gen. Leonard Perroots, USAF Ret., *Signal*, March 1990.
131 Lists of non-lethal technologies: "Nonlethality: Development of a National Policy and Employing Nonlethal Means in a New Strategic Era," a U.S. Global Strategy Council paper, n.d.
132 Doctrinal thinking about non-lethality: "Military Studies Unusual Arsenal," by Neil Munro and Barbara Opall, *Defense News*, October 19–25, 1992.
133 Secrecy and non-lethality: Interviews with Morrises.
134 Diplomacy and non-lethality: Interviews with Morrises.

CHAPTER 16   THE KNOWLEDGE WARRIORS

139 Strassmann: The authors have known Strassmann for many years. This material is largely based on interviews with him.

140   Strassman on lack of information doctrine: "DoD Creates Information Doctrine," by Neil Munro, *Defense News*, December 2, 1991.
140   Net assessment unit: Meeting with Andy Marshall and staff.
140   Knowledge as strategic asset: "Pentagon Wartime Plan Calls for Deception, Electronics," by Neil Munro, *Defense News*, May 10–16, 1993.
141   Knowledge warfare overview: "Cyberwar Is Coming," by John Arquilla and David Ronfeldt, Draft Discussion Paper, RAND International Policy Department, June 1992.
143   Silicon Valley "secret": "ASAP Interview/Tom Peters," *Forbes ASAP*, March 29, 1993.
144   On "connectivity": [407], p. 559.
145   Integrated global network: Interview with Stuart Slade of Forecast International.
146   Political significance of integrated military communications: Interview with Slade.
148   Information superiority as fragile: Interview with Munro.
149   Computer and telecom vulnerability: "Exposure to 'Virus' Is Widespread Among U.S. Funds Transfer Systems," by Steven Mufson, and "FBI Investigates Computer Tapping in Sprint Contract," by John Burgess, both in *International Herald Tribune*, February 22, 1990; "New York Business Warned over Threat of Telecoms Failure," *Financial Times*, June 19, 1990; "U.S. Boosts Information Warfare Initiatives," by Neil Munro, *Defense News*, January 25–31, 1993; "Stealth Virus Attacks," by John Dehaven, *Byte*, May 1993.

     See also: "Soft Kill," by Peter Black, *Wired*, July/August 1993, in which he points out "computer emergency response teams" have reportedly been set up at the Department of Defense, the Department of Energy, and the National Security Agency to deal with viral and other attacks on their computer systems, but that "these are tiny digital fire departments" and "there is no grand strategy for defense or offense" with respect to the information infrastructure.
151   Viral predators, etc.: [251], pp. 126–33.

CHAPTER 17   THE FUTURE OF THE SPY

See also [379], Chap. 24, "A Market for Spies."
153   Kalugin background: [10], pp. 483–84, 525–27.
154   Russian "Academy for State Security": Kalugin interview.
154   "Pretty well covered" editorial: *New York Times*, March 18, 1993.
155   New intelligence "products": "Staying in the National Security Business: New Roles for the U.S. Military," by John L. Petersen, Proceedings of First Symposium of Open Source Solutions, Inc., Washington, D.C., December 1–3, 1992, Vol. I.
156   "Point of sale" intelligence: "Intelligence in the Year 2002: A Concept of Operations," by Andrew Shepard, Proceedings of First Symposium of Open Source Solutions, Inc., Washington, D.C., December 1–3, 1992, Vol. I.
156   Codevilla plan: "The CIA, Losing Its Smarts," by Angelo Codevilla, *New York Times*, February 13, 1993.
156   Wheat and chaff: Problems of evaluating the effectiveness of intelligence are discussed in "Intelligence and U.S. Foreign Policy: How to Measure Success?," by Glenn Hastedt, *International Journal of Intelligence and CounterIntelligence*, Spring 1991, pp. 49–62.

157   Changing requirements for intelligence: [26], p. 190.
157   Precision personal intelligence: [90], p. 137.
157   Zeroing in on terrorists: "Visualizing Patterns and Trends in Data," by Christopher Westphal and Robert Beckman, Proceedings of Symposium on Advanced Information Processing and Analysis Steering Group (Intelligence Community), Tysons Corner, Virginia, March 2–4, 1993.
157   The Analytic Sciences Corporation (TASC) on tracking arms sales: Ibid.
160   Costs of secrecy: Keyworth in Proceedings of First Symposium of Open Source Solutions, Inc., Washington, D.C., December 1–3, 1992, Vol. I.
160   Steele material based on interviews with him as well as the following articles by him: "Applying the 'New paradigm': To Avoid Strategic Failures in the Future," *American Intelligence Journal,* Autumn 1991; "E3I: Ethics, Ecology, Evolution and Intelligence," *Whole Earth Review,* Fall 1992; also numerous papers in "Intelligence — Selected Readings — Book One" of the Command and Staff College, U.S. Marine Corps, Marine Corps University. Also: "Welcoming Remarks" at First Symposium of Open Source Solutions, Inc., Washington, D.C., December 1–3, 1992, Vol. I.

CHAPTER 18   SPIN

166   Ancient Greek propaganda: [371], p. 31.
166   Dezinformatsia: [345].
166   German medal: [371], p. 165.
167   29 million leaflets: [407], p. 537.
167   The Prussian "Ogre": [371], p. 166.
168   Demonization: [372], pp. 6–7, 140, 211.
168   God on our side: [240], p. 102.
170   Six weeks of TV: [349], p. 123.
170   TV seized power: "L'ere du Soupçon," by Ignacio Ramonet, *Le Monde Diplomatique,* May 1991.
172   Battle of New Orleans: [354], pp. 220–21.
173   Media-tization: "La Guerre du Golfe n'a pas en Lieu!" in *Le Matin du Sahara,* June 24, 1991.

CHAPTER 19   PLOUGHSHARES INTO SWORDS

179   Conscription in French Revolution: [289], pp. 10–11.
180   Prussia imitates French war-form: [136], p. 25.
180   Electronic war translates . . . : " 'Grandes Oreilles' contre 'cerveaux,' " *Le Monde,* June 1, 1992.
181   Mortar tubes: Personal communication, May 11, 1993.
184   Lockheed and Livermore conversion ventures: "The Big Switch," by Peter Grier, *World Monitor,* January 1993.
185   Consumer services for war: Interviews with Daniel Goure.
188   Pattern recognition: "The Defense Whizzes Making It in Civvies," *Business Week,* September 7, 1992.
188   Desktop lathe: "Fetish," *Wired,* May–June 1993.
188   Rapid prototyping at Baxter: "Slicing and Molding by Computer," by John Holusha, *New York Times,* April 7, 1993.

CHAPTER 20   THE GENIE UNLEASHED

190   The simulation of a nuclear crisis involving the U.S. and North Korea was designed to be both chastening and instructive. It forces players to con-

sider many of the non-obvious moral, political, and technical questions that would face decision makers in the event of a real crisis.

192   Strategic missiles and warheads in ex-Soviet republics at writing are as follows:

RUSSIA
SS-11 SEGO   280 missiles, 560 warheads (approx.)
SS-13 SAVAGE   40 missiles, 40 warheads
SS-17 SPANKER   40 missiles, 160 warheads
SS-18 SATAN   204 missiles, 2,040 warheads
SS-19 STILETTO   170 missiles, 1,020 warheads
SS-24 SCALPEL   36 missiles, 360 warheads (rail)
SS-24 SCALPEL   10 missiles, 100 warheads
SS-25 SICKLE   260+ missiles, 260+ warheads

UKRAINE
SS-19 STILETTO   130 missiles, 780 warheads
SS-24 SCALPEL   46 missiles, 460 warheads

KAZAKHSTAN
SS-18 SATAN   104 missiles, 1,040 warheads

BELARUS
SS-25 SICKLE   80 missiles, 80 warheads

Looking at this list should lead thoughtful people to consider what could happen if other nuclear-armed nations were to prove fissionable. What happens to France's *force de frappe* if ultra-nationalists were to gain power in Paris at some future date or if separatist movements were to tear France apart? Who would seize China's nukes if civil war were to erupt a decade after Deng Xiaoping's death? For that matter, what about the greatest nuclear power of all, the United States? Could one ever imagine Idaho, home of strategic missile silos and flourishing neo-Nazi cults, attempting someday to break away from so-called domination by Washington? Extremely unlikely. But it was equally difficult at one time to imagine the independence of the Ukraine, Belarus, or Kazakhstan, the Soviet Union's own nuclear-equipped "Idaho."

194   Nukes stored in Russian rail cars: "Parliament Agrees to Slash Weapons Stockpile," by Alexander Stukalin, *Commersant* (Moscow), November 10, 1992. Dangerous conditions also emphasized in interview with Viktor Alksnis, the so-called "Dark Colonel," former leader of Soyuz group on Soviet parliament.

194   On smuggling of Russian nuclear materials, see "It's Time to Stop Russia's Nuclear Mafia," by Kenneth R. Timmerman, *Wall Street Journal,* November 27–28, 1992. See also: "Smuggler's Paradise," by Steve Liesman, *Moscow Times,* December 5–6, 1992.

194   People's Mujahedeen on Kazakhstan sale of nukes to Iran: "It's Time to Stop Russia's Nuclear Mafia," by Kenneth R. Timmerman, *Wall Street Journal,* November 27–28, 1992. Also, "Iran-Kazakhstan Nuclear Deal Stories Denied," *San Jose Mercury News,* October 16, 1992. President Nazarbayev dismissed such "rumors" in a lengthy meeting with us in Alma Ata on December 3, 1992.

194   Azerbaijan nukes: "Osetia amenaza a Georgia con lanzar un ataque nuclear," *ABC* (Barcelona), June 2, 1992.

195   NPT the most widely adhered to treaty: "Iraq and the Future of Nuclear

Nonproliferation: The Roles of Inspections and Treaties," by Joseph F. Pilat, *Science*, March 6, 1992.

196   Iraq N-program "at zero": "Iraq's Bomb — an Update," by Diana Edensword and Fary Milhollin, *New York Times*, April 26, 1993.

196   IAEA inspectors: The Annual Report for 1990, International Atomic Energy Agency, July 1991; also "The Nuclear Epidemic," *U.S. News & World Report*, March 16, 1992.

196   Channels of trade and smuggling: "Smuggler's Paradise," by Steve Liesman, *Moscow Times*, December 5–6, 1992.

196   Interview with Mikhailov on November 27, 1992, in Moscow.

196   Mokhov on N-material thefts: "Ex-Soviets 'Loose Nukes' Sparking Security Concerns," by John-Thor Dahlberg, *Los Angeles Times*, December 28, 1992.

197   Builder material: Interview with Builder; also *The Future of Nuclear Deterrence*, by Carl H. Builder, RAND Paper P-7702, RAND Corporation, February 1991.

198   Export controls uncoordinated: "Iraq's Bomb — an Update," by Diana Edensword and Gary Milhollin, *New York Times*, April 26, 1993. Also interview with Edensword.

201   Reconceptualization of proliferation problem: Interviews with Seaquist. Also a working paper of the Counter-Proliferation Initiative in the Office of the U.S. Secretary of Defense.

202   Golay: Cited in "The Nuclear Epidemic," *U.S. News & World Report*, March 16, 1992. For a scorching attack on release of nuclear information, see also: "Proliferation 101: The Presidential Faculty," by Arnold Kramish, *Global Affairs*, Spring 1993.

202   The flow of information: *The Future of Nuclear Deterrence*, by Carl H. Builder, RAND Paper P-7702, RAND Corporation, February 1991.

If we look at the nuclear menace not as a short-term phenomenon, but as a 25–30-year problem, it suggests the need for long-range work on technologies to neutralize or at least reduce the danger. We need better technical means to detect radioactivity — even if shielded or buried. We know that Electro-Magnetic Pulse (EMP) can be produced by non-nuclear means and can essentially fry the electronics on which nuclear arms depend. EMP weaponry should be high on the research agenda. We need better robots to help protect existing nuclear facilities from terrorists, criminals, and others who might seek to enter or damage them. We need better and safer permissive action links . . . more sensitive sensors . . . better satellite imagery and data fusion . . . and more and more precision in alternate weaponry. In short, the knowledge-intensive technologies can help reduce the threat from nuclear weapons in the world.

There is no sure protection against maniacs bent on revenge or collective suicide, but Third Wave tools are needed to neutralize the ultimate weapon of the Second Wave.

CHAPTER 21   THE ZONE OF ILLUSION

207   Technological base for regionalism: Japan's Kenichi Ohmae has tracked the rise of the region-state and described the nation-state as "dysfunctional." In the Spring 1993 issue of *Foreign Affairs*, he pointed out that the region-states' "primary linkage" is with "the global economy and not with their host nations." But Ohmae assumes that "traditional issues of foreign policy, security and defense" along with macroeconomic and monetary policies will remain the "province of nation-states." He urges nation-states

to treat the rising power of region-states "gently" and scants, in the *Foreign Affairs* piece, the political implications of bi- and even tri-national regions. Ohmae is one of the smartest global analysts we have, but we believe he underestimates the political earthquake that the rise of regional power is likely to trigger.

Rising regions will not allow nation-states indefinitely to set their taxes, decide their trade policies, manipulate their currency, and represent them diplomatically. (The very same issue of *Foreign Affairs* features an article calling on California to adopt its own foreign policy.) Regions must inevitably challenge national power, and when they do there is no reason to assume central authorities will treat them "gently." Moreover, the rise of the region is not just a matter of economic rationality — it involves culture, religion, ethnicity, and other deeply emotional and, hence, politically dangerous conflicts.

209   Re Electronic-Political Networks: "Electronic Democracy," by Howard H. Frederick, *Edges* (Toronto), July–September 1992.

CHAPTER 22   A WORLD TRISECTED

214   China's swollen bellies: "As China Leaps Ahead, the Poor Slip Behind," by Sheryl WuDunn, *New York Times,* May 23, 1993.
215   India's Lajpat-Rai market: "Dish-Wallahs," by Jeff Greenwald, *Wired,* May–June 1993.
215   Separatists in Brazil: "Trying to Head Off a Brazilian Breakaway," by Christina Lamb, *Financial Times,* November 3, 1992.

CHAPTER 23   ABOUT PEACE-FORMS

223   Primitive attempts to mitigate violence: [86], pp. 176–79.
224   Norms for treating combatants: [2], pp. 27–30.

CHAPTER 24   THE NEXT PEACE-FORM

226   Ideas about peace unchanged since 1815: [23], p. v.
230   Open skies: [46], pp. 26–27.
231   Acceptance of inspection: "Future of Monitoring and Verification," by Hendrik Wagenmakers, paper submitted to UN Conference on "A Post-Cold War International System and Challenges to Multilateral Disarmament Efforts," Kyoto, Japan, May 27–30, 1991.
232   IAEA failures: "Iraqi Atom Effort Exposes Weakness in World Controls," by William J. Broad, *New York Times,* July 15, 1991.
234   Meshad and Bull assassinations: [1], pp. xiii, 18.
235   Permissive Action Links: "Star Wars in the Twilight Zone," *New York Times,* June 14, 1992.
236   Vasic: "Quiet Voices from the Balkans," *The New Yorker,* March 3, 1993.
237   Orosi: "Albanian Journalism: First Victim of the Media War," by Violeta Orosi in *Pristina,* reprinted in *War Report* (London), April/May 1993.
237   Peace Action: Interview with Aaron.
238   U.S. radio: "U.S. Plans Radio Free Serbia in Bid to Weaken Milosevic," by Doyle McManus, *Los Angeles Times,* June 21, 1993.
238   Dame Melba Sings: [409], p. 176.
239   Digital revolution: "Information Revolutions and the End of History," by Elin Whitney-Smith, paper submitted to symposium of Open Source Solutions, Inc., Washington, D.C., December 1–3, 1992.

CHAPTER 25   THE NEW GLOBAL SYSTEM

242   5,000 countries: "As Ethnic Wars Multiply, U.S. Strives for a Policy," by David Binder and Barbara Crossette, *New York Times*, February 7, 1993.

242   Singapore city-states: Interview with Yao.

243   Technopolis: "Techno-Apartheid for a Global Underclass," by Riccardo Petrella, *Los Angeles Times*, August 6, 1992.

244   500 sub-companies: "Inside Unilever: The Evolving Transnational Company," by Floris A. Maljers, *Business Review*, September–October, 1992.

244   AT&T and UN figures: "Global Link-up Down the Line," by Andrew Adonis, *Financial Times*, June 5, 1993.

245   Dresden skinheads: "Electronic Democracy," by Howard H. Frederick, *Edges* (Toronto), July–September 1992.

250   Prigogine on non-equilibrium: [300].

BIBLIOGRAPHY

[1] Adams, James. *Bull's Eye*. (New York: Times Books, 1992.)

[2] Ahlstrom, Christer, and Kjell-Ake Nordquist. *Casualties of Conflict*. (Sweden: Uppsala University, 1991.)

[3] Al-Khalil, Samir. *Republic of Fear: The Politics of Modern Iraq*. (Berkeley, CA: University of California Press, 1989.)

[4] Aldridge, Robert C. *The Counterforce Syndrome*. (Washington, D.C.: Institute for Policy Studies, 1981.)

[5] Alexander, Yonah, Y. Ne'eman, and E. Tavin. *Terrorism*. (Washington, D.C.: Global Affairs, 1991.)

[6] Alpher, Joseph, ed. *War in the Gulf: Implications for Israel*. (Jerusalem: Jaffee Center Study Group, 1992.)

[7] Amalrik, Andrei. *Will the Soviet Union Survive Until 1984?* (New York: Perennial Library, 1970.)

[8] Andrew, Christopher. *Secret Service*. (London: William Heinemann, 1985.)

[9] Andrew, Christopher, and David Dilks, eds. *The Missing Dimension*. (Urbana: University of Illinois Press, 1985.)

[10] Andrew, Christopher, and Oleg Gordievsky. *KGB: The Inside Story*. (New York: HarperPerennial, 1990.)

[11] Arkin, William M., J. M. Handler, J. A. Morrissey, and J. M. Walsh. *Encyclopedia of the U.S. Military*. (New York: Harper & Row, 1990.)

[12] Aron, Raymond. *On War*. (New York: W. W. Norton, 1968.)

[13] Arquilla, John. *Dubious Battles*. (Washington, D.C.: Crane Russak, 1992.)

[14] Asprey, Robert B. *War in the Shadows: Vol. I and II*. (Garden City, NY: Doubleday, 1975.)

[15] Bailey, Kathleen C. *Doomsday Weapons in the Hands of Many*. (Chicago: University of Illinois Press, 1991.)

[16] Baker, David. *The Shape of Wars to Come*. (Cambridge, England: Patrick Stephens, 1981.)

[17] Bamford, James. *The Puzzle Palace*. (Boston: Houghton Mifflin, 1982.)

[18] Barcelona, Eduardo, and Julio Villalonga. *Relaciones Carnales*. (Buenos Aires: Planeta, 1992.)

[19]   Barnet, Richard J. *Roots of War.* (Baltimore: Penguin, 1973.)

[20]   Barringer, Richard E. *War: Patterns of Conflict.* (Cambridge, MA: The MIT Press, 1972.)

[21]   Baxter, William P. *Soviet Airland Battle Tactics.* (Novato, CA: Presidio Press, 1986.)

[22]   Baynes, J. C. M. *The Soldier and Modern Society.* (London: Eyre Methuen, 1972.)

[23]   Beales, A. C. F. *The History of Peace.* (London: G. Bell and Sons, 1931.)

[24]   Beaumont, Roger A. *Military Elites.* (New York: Bobbs-Merrill, 1974.)

[25]   Beckwith, Charlie A., and Donald Knox. *Delta Force.* (New York: Harcourt Brace Jovanovich, 1983.)

[26]   Berkowitz, Bruce D., and Allan E. Goodman. *Strategic Intelligence.* (Princeton, NJ: Princeton University Press, 1991.)

[27]   Best, Geoffrey. *War and Society in Revolutionary Europe: 1770–1870.* (Leicester, England: Fontana, 1982.)

[28]   Bibo, Istvan. *The Paralysis of International Institutions and the Remedies.* (New York: John Wiley & Sons, 1976.)

[29]   Bidwell, Shelford, ed. *World War 3.* (Feltham, England: Hamlyn Paperbacks, 1979.)

[30]   Bienen, Henry. *Violence and Social Change.* (Chicago: The University of Chicago Press, 1970.)

[31]   ——, ed. *The Military Intervenes.* (Hartford, CT: Russell Sage Foundation, 1968.)

[32]   Blackwell, James. *Thunder in the Desert.* (New York: Bantam, 1991.)

[33]   Blechman, Barry M., et al. *Force Without War.* (Washington, D.C.: The Brookings Institution, 1978.)

[34]   Bloomfield, Lincoln P., and Amelia C. Leiss. *Controlling Small Wars.* (New York: Knopf, 1969.)

[35]   Booth, Ken. *Strategy and Ethnocentrism.* (London: Croon Helm, 1979.)

[36]   ——, ed. *New Thinking About Strategy and International Security.* (London: Harper-Collins Academic, 1991.)

[37]   Boserup, Anders, and Andrew Mack. *War Without Weapons.* (London: Frances Pinter, 1974.)

[38]   Boulding, Kenneth. *The Meaning of the Twentieth Century.* (New York: Harper, 1964).

[39]   Braddon, Russell. *Japan Against the World: 1941–2041.* (New York: Stein and Day, 1983.)

[40]   Brandon, David H., and Michael A. Harrison. *The Technology War.* (New York: John Wiley & Sons, 1987.)

[41]   Braudel, Fernand. *The Mediterranean.* (New York: Harper & Row, 1973.)

[42]   ——. *The Structures of Everyday Life.* (New York: Harper & Row, 1979.)

[43]   Brockway, Fenner, and Frederic Mullally. *Death Pays a Dividend.* (London: Victor Gollancz, 1944.)

[44]   Brodie, Bernard, and Fawn M. Brodie. *From Crossbow to H-Bomb.* (Bloomington: Indiana University Press, 1973.)

[45]   Bruce-Briggs, B. *The Shield of Faith.* (New York: Simon and Schuster, 1988.)

[46] Brugioni, Dino A. *Eyeball to Eyeball*. (New York: Random House, 1991.)

[47] Brzezinski, Zbigniew. *Out of Control*. (New York: Charles Scribner's Sons, 1993.)

[48] Buchanan, Allen. *Secession*. (Boulder, CO: Westview Press, 1991.)

[49] Builder, Carl H. *The Future of Nuclear Deterrence, P-7702*. (Santa Monica, CA: The RAND Corporation, 1990.)

[50] Burr, John G. *The Framework of Battle*. (New York: J. B. Lippincott, 1943.)

[51] Burrows, William E. *Deep Black*. (New York: Random House, 1986.)

[52] Burton, Anthony. *Revolutionary Violence*. (New York: Crane, Russak, 1978.)

[53] Campen, Alan D., ed. *The First Information War*. (Fairfax, VA: AFCEA International Press, 1992.)

[54] Carlton, David, and Carlo Schaerf, eds. *International Terrorism and World Security*. (London: Croom Helm, 1975.)

[55] Carr, Harry. *Riding the Tiger*. (Cambridge, MA: Riverside Press, 1934.)

[56] Chace, James. *The Consequences of the Peace*. (New York: Oxford University Press, 1992.)

[57] Chakotin, Serge. *The Rape of the Masses*. (New York: Alliance, 1940.)

[58] Chatfield, Charles, ed. *Peace Movements in America*. (NewYork: Schocken, 1973.)

[59] Cipolla, Carlo M. *Before the Industrial Revolution*. (New York: W. W. Norton, 1976.)

[60] Clark, Doug. *The Coming Oil War*. (Irvine, CA: Harvest House, 1980.)

[61] Clark, George. *Early Modern Europe*. (New York: Galaxy, 1960.)

[62] Clarke, I. F. *Voices Prophesying War: 1763–1984*. (New York: Oxford University Press, 1966.)

[63] Clausewitz, Carl von. *On War*. (New York: Viking Penguin, 1988.)

[64] ———. *On War*. (Washington, D.C.: Infantry Journal Press, 1950.)

[65] ———. *Principles of War*. (Harrisburg, PA: Stackpole, 1960.)

[66] Clayton, James L. *Does Defense Beggar Welfare?* (New York: National Strategy Information Center, 1979.)

[67] Clutterbuck, Richard. *Kidnap and Ransom: The Response*. (Boston: Faber and Faber, 1978.)

[68] Cohen, Eliot A., and John Gooch. *Military Misfortunes: The Anatomy of Failure in War*. (New York: Vintage, 1991.)

[69] Cohen, Sam. *The Truth About the Neutron Bomb*. (New York: William Morrow, 1983.)

[70] Colby, Charles C., ed. *Geographic Aspects of International Relations*. (Port Washington, NY: Kennikat Press, 1970.)

[71] Coleman, J. D. *Incursion*. (New York: St. Martin's, 1991.)

[72] Collins, John M. *Military Space Forces: The Next 50 Years*. (Washington, D.C.: Pergamon-Brasseys, 1989.)

[73] Cordesman, Anthony, and Abraham Wagner. *Lessons of Modern War: The Arab-Israeli Conflicts, 1973–1988, Vol. I*. (Boulder, CO: Westview Press, 1990.)

[74] Corvisier, Andre. *Armies and Societies in Europe: 1494–1789*. (Bloomington: Indiana University Press, 1979.)

[75]   Crankshaw, Edward. *The Fall of the House of Hapsburg.* (New York: Penguin, 1983.)

[76]   Crenshaw, Martha, ed. *Terrorism, Legitimacy, and Power.* (Middletown, CT: Wesleyan University Press, 1983.)

[77]   Creveld, Martin Van. *Command in War.* (Cambridge, MA: Harvard University Press, 1985.)

[78]   ———. *Supplying War.* (New York: Cambridge University Press, 1977.)

[79]   Croix, Horst De La. *Military Considerations in City Planning: Fortifications.* (New York: George Braziller, 1972.)

[80]   Cross, James Eliot. *Conflict in the Shadows: The Nature and Politics of Guerilla War.* (Garden City, NY: Doubleday, 1963.)

[81]   Crozier, Brian. *A Theory of Conflict.* (London: Hamish Hamilton, 1974.)

[82]   Cunliffe, Marcus. *The Age of Expansion: 1847–1917.* (Springfield, MA: G. & C. Merriman, 1974.)

[83]   Curtin, Philip D., ed. *Imperialism.* (New York: Walker, 1971.)

[84]   D'Albion, Jean. *Une France sans Defense.* (Lonrai, France: Calmann-Levy, 1991.)

[85]   Davidow, William H., and Michael S. Malone. *The Virtual Corporation.* (New York: HarperBusiness, 1992.)

[86]   Davie, Maurice R. *The Evolution of War.* (New Haven, CT: Yale University Press, 1929.)

[87]   de Gaulle, Charles. *The Edge of the Sword.* (Westport, CT: Greenwood Press, 1960.)

[88]   de Jouvenal, Bertrand. *On Power.* (Boston: Beacon Press, 1969.)

[89]   de Lupos, Ingrid Detter. *The Law of War.* (New York: Cambridge University Press, 1987.)

[90]   de Marenches, Count, and David A. Andelman. *The Fourth World War.* (New York: William Morrow, 1992.)

[91]   de Marenches, Count, and Christine Ockrent. *The Evil Empire.* (London: Sidgwick & Jackson, 1988.)

[92]   de Seversky, Maj. Alexander P. *Victory Through Airpower.* (New York: Simon and Schuster, 1942.)

[93]   Deacon, Richard. *A History of the Russian Secret Service.* (London: Frederick Muller, 1972.)

[94]   ———. *The French Secret Service.* (London: Grafton, 1990.)

[95]   Delbruck, Hans. *The Barbarian Invasions: History of the Art of War, Vol. II.* (Lincoln: University of Nebraska Press, 1990.)

[96]   ———. *Medieval Warfare: History of the Art of War, Vol. III.* (Lincoln: University of Nebraska Press, 1990.)

[97]   ———. *Warfare in Antiquity: History of the Art of War, Vol. I.* (Lincoln: University of Nebraska Press, 1990.)

[98]   Derrer, Douglas S. *We Are All the Target.* (Annapolis, MD: Naval Institute Press, 1992.)

[99]   Diagram Group, ed. *Weapons.* (New York: St. Martin's, 1990.)

[100] Dolgopolov, Yevgeny. *The Army and the Revolutionary Transformation of Society.* (Moscow: Progress, 1981.)

[101] Donovan, James A. *U.S. Military Force — 1980: An Evaluation.* (Washington, D.C.: Center for Defense Information, 1980.)

[102] Douhet, Giulio. *The Command of the Air.* (New York: Coward-McCann, 1942.)

[103] Dower, John W. *War Without Mercy.* (New York: Pantheon, 1986.)

[104] Drexler, Eric, and Chris Peterson with Gayle Pergamit. *Unbounding the Future.* (New York: William Morrow, 1991.)

[105] Drucker, Peter F. *Post-Capitalist Society.* (New York: HarperBusiness, 1993.)

[106] Dunn, Richard S. *The Age of Religious Wars: 1559–1715.* (New York: W. W. Norton, 1979.)

[107] Dupuy, Col. T. N. *The Evolution of Weapons and Warfare.* (London: Jane's, 1980.)

[108] ———. *Numbers, Predictions & War.* (New York: Bobbs-Merrill, 1979.)

[109] ———. *Understanding War.* (New York: Paragon House, 1987.)

[110] Duyvendak, J. J. L., trans. *The Book of Lord Shang.* (London: Arthur Probsthain, 1963.)

[111] Earle, Edward Meade, ed. *Makers of Modern Strategy.* (Princeton, NJ: Princeton University Press, 1973.)

[112] Edgerton, Robert B. *Sick Societies.* (New York: The Free Press, 1992.)

[113] Ellis, John. *The Social History of the Machine Gun.* (New York: Pantheon, 1975.)

[114] Fawcett, J. E. S. *The Law of Nations.* (New York: Basic Books, 1968.)

[115] Ferrill, Arthur. *The Origins of War.* (London: Thames & Hudson, 1988.)

[116] Finer, S. E. *The Man on Horseback: The Role of the Military in Politics.* (London: Pall Mall Press, 1969.)

[117] Fletcher, Raymond. *60 Pounds a Second on Defence.* (London: MacGibbon & Kee, 1963.)

[118] Ford, Daniel. *The Button.* (New York: Simon and Schuster, 1985.)

[119] Franck, Thomas M., and Edward Weisband. *Secrecy and Foreign Policy.* (New York: Oxford University Press, 1974.)

[120] Fromkin, David. *A Peace to End All Peace.* (New York: Avon, 1990.)

[121] Fukuyama, Francis. *The End of History and the Last Man.* (New York: Avon, 1992.)

[122] Galbraith, John Kenneth. *How to Control the Military.* (Garden City, NY: Doubleday, 1969.)

[123] Gallagher, James J. *Low-Intensity Conflict.* (Harrisburg, PA: Stackpole Books, 1992.)

[124] Gallois, Pierre M. *Geopolitique les Voies de la Puissance.* (Paris: Fondation des Etudes de Defense Nationale, 1990.)

[125] Gasparini Alves, Pericles. *The Interest of Nonpossessor Nations in the Draft Chemical Weapons Convention.* (New York: Vantage, 1990.)

[126] Geary, Conor. *Terror.* (London: Faber and Faber, 1991.)

[127] Geraghty, Tony. *Inside the S.A.S.* (New York: Ballantine, 1982.)

[128] Gerard, Francis. *Vers L'unite Federale du Monde.* (Paris: Denoel, 1971.)

[129] Gervasi, Tom. *Arsenal of Democracy.* (New York: Grove, 1977.)

[130] Geyer, Alan. *The Idea of Disarmament!* (Elgin, IL: The Brethren Press, 1982.)

[131] Giap, Vo Nguyen. *Banner of the People's War, the Party's Military Line.* (New York: Praeger, 1970.)

[132]   Gilpin, Robert. *War and Change in World Politics*. (New York: Cambridge University Press, 1985.)

[133]   Ginsberg, Robert, ed. *The Critique of War*. (Chicago: Henry Regnery, 1970.)

[134]   Godson, Roy. *Intelligence Requirements for the 1980's: Domestic Intelligence*. (Lexington, MA: Lexington, 1986.)

[135]   Goerlitz, Walter. *History of the German General Staff: 1657–1945*. (New York: Praeger, 1956.)

[136]   Gooch, John. *Armies in Europe*. (London: Routledge & Kegan Paul, 1980.)

[137]   Goodenough, Simon. *Tactical Genius in Battle*. (New York: E. P. Dutton, 1979.)

[138]   Grant, Michael. *A History of Rome*. (New York: Scribner, 1978.)

[139]   Gray, Colin S. *House of Cards*. (Ithaca, NY: Cornell University Press, 1992.)

[140]   Hackett, John. *The Third World War: The Untold Story*. (New York: Bantam, 1983.)

[141]   Halamka, John D. *Espionage in Silicon Valley*. (Berkeley, CA: Sybex, 1984.)

[142]   Hale, J. R. *Renaissance Europe, 1480–1520*. (London: Collins, 1971.)

[143]   Halperin, Morton H. *Contemporary Military Strategy*. (Boston: Little, Brown, 1967.)

[144]   Hanson, Victor Davis. *The Western Way of War*. (New York: Oxford University Press, 1989.)

[145]   Harries, Meirion, and Susie Harries. *Soldiers of the Sun*. (New York: Random House, 1991.)

[146]   Hart, B. H. Liddell. *Europe in Arms*. (London: Faber and Faber, 1937.)

[147]   ———. *Strategy*. (New York: Meridien, 1991.)

[148]   Hartigan, Richard Shelly. *The Forgotten Victim: A History of the Civilian*. (Chicago: Precedent, 1982.)

[149]   Hartogs, Renatus, and Eric Artzt. *Violence: Causes & Solutions*. (New York: Dell, 1970.)

[150]   Herzog, Chaim. *The Arab-Israeli Wars*. (New York: Random House, 1982.)

[151]   Hill, Christopher. Reformation to Industrial Revolution: 1530–1780. (Baltimore: Penguin Books, 1969).

[152]   Hobsbawm, E. J. *Industry and Empire*. (Baltimore, MD: Penguin, 1969.)

[153]   Hoe, Alan. *David Stirling*. (London: Little, Brown, 1992.)

[154]   Hofstadter, Richard, William Miller, and Daniel Aaron. *The United States*. (Englewood Cliffs, NJ: Prentice-Hall, 1967.)

[155]   Hohne, Heinze, and Hermann Zolling. *The General Was a Spy*. (New York: Coward, McCann & Geoghegan, 1972.)

[156]   Holsti, Kalevi J. *Peace and War: Armed Conflicts and International Order, 1648–1989*. (Cambridge, England: Cambridge University Press, 1991.)

[157]   Honan, William H. *Bywater: The Man Who Invented the Pacific War*. (London: Macdonald, 1990.)

[158]   Hoselitz, Bert F., and Wilbert E. Moore. *Industrialization*. (n.p.: UNESCO-Mouton, 1968.)

[159]   Howard, Michael. *The Causes of Wars*. (London: Unwin Paperbacks, 1983.)

[160]   ———. *War and the Liberal Conscience*. (New Brunswick, NJ: Rutgers University Press, 1986.)

[161]   ———. *War in European History.* (New York: Oxford University Press, 1989.)

[162]   Hoyt, Edwin P. *Japan's War.* (New York: McGraw-Hill, 1986.)

[163]   Hughes, Wayne P. *Fleet Tactics.* (Annapolis, MD: Naval Institute Press, 1986.)

[164]   Huie, William Bradford. *The Case Against the Admirals.* (New York: E. P. Dutton, 1946.)

[165]   Huntington, Samuel P. *The Soldier and the State.* (Cambridge, MA: The Belknap Press, 1957.)

[166]   Janowitz, Morris. *The Military in the Political Development of New Nations.* (Chicago: The University of Chicago Press, 1971.)

[167]   ———, ed. *The New Military: Changing Patterns of Organization.* (New York: Russell Sage Foundation, 1964.)

[168]   Johnson, James Turner, and George Weigel. *Just War and the Gulf War.* (Washington, D.C.: Ethics and Public Policy Center, 1991.)

[169]   Jones, Ellen. *Red Army and Society.* (Boston: Allen and Unwin, 1985.)

[170]   Jones, J. *Stealth Technology.* (Blue Ridge Summit, PA: Aero, 1989.)

[171]   Joyce, James Avery. *The War Machine: The Case Against the Arms Race.* (New York: Discus, 1982.)

[172]   Juergensmeyer, Mark. *The New Cold War.* (Berkeley, CA: University of California Press, 1993.)

[173]   Kahalani, Avigdor. *The Heights of Courage.* (New York: Praeger, 1992.)

[174]   Kahan, Jerome H. *Security in the Nuclear Age.* (Washington, D.C.: The Brookings Institution, 1975.)

[175]   Kaldor, Mary. *The Baroque Arsenal.* (New York: Hill and Wang, 1981.)

[176]   Kaplan, Fred. *The Wizards of Armageddon.* (New York: Simon and Schuster, 1983.)

[177]   Katz, Howard S. *The Warmongers.* (New York: Books in Focus, 1981.)

[178]   Kaufmann, William W. *A Thoroughly Efficient Navy.* (Washington, D.C.: The Brookings Institution, 1987.)

[179]   Keith, Arthur Berriedale. *The Causes of War.* (New York: Thomas Nelson and Sons, 1940.)

[180]   Kellner, Douglas. *The Persian Gulf TV War.* (Boulder, CO: Westview Press, 1992.)

[181]   Kennedy, Gavin. *The Military in the Third World.* (London: Duckworth, 1974.)

[182]   Kennedy, Malcolm J., and Michael J. O'Connor. *Safely by Sea.* (Lanham, MD: University Press of America, 1990.)

[183]   Kennedy, Paul. *The Rise and Fall of Great Powers.* (New York: Random House, 1987.)

[184]   ———, ed. *Grand Strategies in War and Peace.* (New Haven, CT: Yale University Press, 1991.)

[185]   Keohane, Robert O., and Joseph S. Nye. *Power and Interdependence.* (Boston: Little, Brown, 1977.)

[186]   Kernan, W. F. *Defense Will Not Win the War.* (Boston: Little, Brown, 1942.)

[187]   Kissin, S. F. *War and the Marxists.* (Boulder, CO: Westview Press, 1989.)

[188] Knightly, Phillip. *The Second Oldest Profession*. (New York: W. W. Norton, 1986.)

[189] Knowles, L. C. A. *The Industrial and Commercial Revolutions in Great Britain during the Nineteenth Century*. (London: George Routledge, 1922).

[190] Kohn, Hans. *The Idea of Nationalism*. (Toronto: Collier, 1944.)

[191] Krader, Lawrence. *Formation of the State*. (Englewood Cliffs, NJ: Prentice-Hall, 1968.)

[192] Kriesel, Melvin E. *Psychological Operations: A Strategic View — Essays on Strategy*. (Washington, D.C.: National Defense University Press, 1985.)

[193] Kull, Irving S., and Nell M. Kull. *The Encyclopedia of American History*. (New York: Popular Library, 1952.)

[194] Kupperman, Robert H., and Darrell M. Trent. *Terrorism: Threat, Reality, Response*. (Stanford, CA: Hoover Institution Press, 1980.)

[195] Laffin, John. *Links of Leadership*. (New York: Abelard-Schuman, 1970.)

[196] Lamont, Lansing. *Day of Trinity*. (New York: Signet, 1966.)

[197] Lang, Walter N. *The World's Elite Forces*. (London: Salamander, 1987.)

[198] Langford, David. *War in 2080: The Future of Military Techology*. (New York: William Morrow, 1979.)

[199] Lansdale, Edward Geary. *In the Midst of Wars: An American's Mission to Southeast Asia*. (New York: Harper & Row, 1972.)

[200] Laqueur, Walter. *Guerrilla*. (London: Weidenfeld & Nicolson, 1977.)

[201] ———. *Terrorism*. (London: Weidenfeld & Nicolson, 1978.)

[202] ———. *A World of Secrets*. (New York: Basic, 1985.)

[203] Latey, Maurice. *Patterns of Tyranny*. (New York: Atheneum, 1969.)

[204] Laulan, Yves Marie. *La Planete Balkanisee*. (Paris: Economica, 1991.)

[205] Laurie, Peter. *Beneath the City Streets*. (London: Granada, 1983.)

[206] Lea, Homer. *The Valor of Ignorance*. (New York: Harper & Brothers, 1909.)

[207] Lederer, Emil. *State of the Masses*. (New York: Howard Fertig, 1967.)

[208] Lenin, V. I. *Lenin on War and Peace*. (Peking: Foreign Language Press, 1966.)

[209] Lentz, Theodore F. *Towards a Science of Peace*. (New York: Bookman Associates, 1961.)

[210] Levite, Ariel. *Intelligence and Strategic Surprises*. (New York: Columbia University Press, 1987.)

[211] Levy, Jack S. *War in the Modern Great Power System: 1495–1975*. (Lexington: University of Kentucky Press, 1983.)

[212] Lewin, Ronald. *Hitler's Mistakes*. (New York: William Morrow, 1984.)

[213] Liebknecht, Karl. *Militarism and Anti-Militarism*. (Cambridge, England: Rivers Press, 1973.)

[214] Lifton, Robert Jay, and Richard Falk. *Indefensible Weapons*. (New York: Basic, 1982.)

[215] Lloyd, Peter C. *Classes, Crises and Coups*. (New York: Praeger, 1972.)

[216] London, Perry. *Behavior Control*. (New York: Harper & Row, 1969.)

[217] Lovell, John P., and Philip S. Kronenberg. *New Civil-Military Relations*. (New Brunswick, NJ: Transaction, 1974.)

[218] Lupinski, Igor. *In the General's House*. (Santa Barbara, CA: Res Gestae Press, 1993.)

[219] Luttwak, Edward. *On the Meaning of Victory.* (New York: Simon and Schuster, 1986.)

[220] ———. *The Pentagon and the Art of War.* (New York: Simon and Schuster, 1984.)

[221] Luttwak, Edward, and Stuart Koehl. *The Dictionary of Modern War.* (New York: HarperCollins, 1991.)

[222] Luvaas, Jay, ed. and trans. *Frederick the Great on the Art of War.* (New York: The Free Press, 1966.)

[223] Machiavelli, Niccolo. *The Art of War.* (New York: Da Capo, 1990.)

[224] Macksey, Kenneth, and William Woodhouse. *The Penguin Encyclopedia of Modern Warfare.* (New York: Viking, 1991.)

[225] Mahan, Alfred T. *Lessons of the War with Spain.* (Freeport, NY: Books for Libraries Press, 1970.)

[226] Mandelbaum, Michael. *The Nuclear Revolution.* (New York: Cambridge University Press, 1981.)

[227] Mansfield, Sue. *The Gestalts of War.* (New York: The Dial Press, 1982.)

[228] Markham, Felix. *Napoleon.* (New York: Mentor, 1963.)

[229] Markov, Walter, ed. *Battles of World History.* (New York: Hippocrene, 1979.)

[230] Maswood, S. Javed. *Japanese Defense.* (Singapore: Institute of Southeast Asian Studies, 1990.)

[231] Maxim, Hudson. *Defenseless America.* (New York: Hearst's International Library Co., 1915.)

[232] Mayer, Arno J. *The Persistence of the Old Regime.* (New York: Pantheon, 1981.)

[233] Mazarr, Michael J. *Missile Defences and Asian-Pacific Security.* (London: Macmillan, 1989.)

[234] McGwire Michael, K. Booth, and J. McDonnell, eds. *Soviet Naval Policy.* (New York: Praeger, 1975.)

[235] McMaster, R. E., Jr. *Cycles of War.* (Kalispell, MT: Timberline Trust, 1978.)

[236] McNeill, William. *The Pursuit of Power.* (Chicago: The University of Chicago Press, 1982.)

[237] Melvern, Linda, D. Hebditch, and N. Anning. *Techno-Bandits.* (Boston: Houghton Mifflin, 1984.)

[238] Mendelssohn, Kurt. *The Secret of Western Domination.* (New York: Praeger, 1976.)

[239] Merleau-Ponty, Maurice. *Humanism and Terror.* (Boston: Beacon Press, 1969.)

[240] Mernissi, Fatima. *Islam and Democracy.* (Reading, MA: Addison-Wesley, 1992.)

[241] Merton, Thomas, ed. *Gandhi on Non-Violence.* (New York: New Directions, 1965.)

[242] Meyer, Cord. *Facing Reality: From World Federalism to the CIA.* (New York: Harper & Row, 1980.)

[243] Miller, Abraham H. *Terrorism and Hostage Negotiations.* (Boulder, CO: Westview Press, 1980.)

[244] Miller, Judith, and Laurie Mylroie. *Saddam Hussein and the Crisis in the Gulf.* (New York: Times Books, 1990.)

[245] Millis, Walter. *Arms and Men.* (New York: Mentor, 1958.)

[246]   ———. *The Martial Spirit.* (Cambridge, MA: The Riverside Press, 1931.)

[247]   Mills, C. Wright. *The Causes of World War Three.* (London: Camelot Press, 1959.)

[248]   Minc, Alain. *La Vengeance des Nations.* (Paris: Bernard Grasset, 1990.)

[249]   Mirsky, Jeanette, and Allan Nevins. *The World of Eli Whitney.* (New York: Macmillan, 1952.)

[250]   Mische, Gerald, and Patricia Mische. *Toward a Human World Order.* (New York: Paulist Press, 1977.)

[251]   Moravec, Hans. *Mind Children.* (Cambridge, MA: Harvard University Press, 1988.)

[252]   Morison, Samuel Eliot. *American Contributions to the Strategy of World War II.* (London: Oxford University Press, 1958.)

[253]   Moro, D. Ruben. *Historia del Conflicto del Atlantico Sur.* (Buenos Aires: Fuerza Aerea Argentina, 1985.)

[254]   Moss, Robert. *The War for the Cities.* (New York: Coward, McCann & Geoghegan, 1972.)

[255]   Motley, James Berry. *Beyond the Soviet Threat.* (Lexington, MA: Lexington, 1991.)

[256]   Mueller, John. *Retreat from Doomsday: The Obsolescence of Major War.* (New York: Basic, 1990.)

[257]   Munro, Neil. *The Quick and the Dead: Electronic Combat and Modern Warfare.* (New York: St. Martin's, 1991.)

[258]   Murphy, Thomas Patrick, ed. *The Holy War.* (Columbus: Ohio State University Press, 1976.)

[259]   Nakdimon, Shlomo. *First Strike.* (New York: Summit, 1987.)

[260]   Naude, Gabriel. *Considerations politiques sur les Coups d'Etat.* (Paris: Editions de Paris, 1988.)

[261]   Nazurbayev, Nursultan. *No Rightists Nor Leftists.* (New York: Noy Publications, 1992.)

[262]   Nelson, Joan M. *Access to Power.* (Princeton, NJ: Princeton University Press, 1979.)

[263]   Nelson, Keith L., and Spencer C. Olin, Jr. *Why War?* (Berkeley, CA: University of California Press, 1980.)

[264]   Netanyahu, Benjamin, ed. *Terrorism.* (New York: Farrar, Straus and Giroux, 1986.)

[265]   Nicholson, Michael. *Conflict Analysis.* (London: The English Universities Press, 1970.)

[266]   Nolan, Keith William. *Into Cambodia.* (New York: Dell, 1991.)

[267]   Nye, Joseph S., Jr. *Bound to Lead.* (New York: Basic, 1990.)

[268]   Nystrom, Anton. *Before, During and After 1914.* (London: William Heinemann, 1915.)

[269]   O'Brien, Conor Cruise. *The Siege: The Saga of Israel and Zionism.* (New York: Simon and Schuster, 1986.)

[270]   Odom, William E. *On Internal War.* (Durham, NC: Duke University Press, 1992.)

[271]   Ohmae, Kenichi. *The Borderless World.* (New York: HarperCollins, 1990.)

[272]   Oppenheimer, Franz. *The State.* (New York: Free Life Editions, 1942.)

[273]   Oren, Nissan, ed. *Termination of Wars*. (Jerusalem: The Magnes Press, 1982.)

[274]   Organski, A. F. K., and Jacek Kluger. *The War Ledger*. (Chicago: The University of Chicago Press, 1980.)

[275]   Osgood, Robert E., and Robert E. Tucker. *Force, Order and Justice*. (Baltimore: Johns Hopkins Press, 1967.)

[276]   Ostrovsky, Victor, and Claire Hoy. *By Way of Deception*. (New York: St. Martin's, 1990.)

[277]   Owen, David Edward. *Imperialism and Nationalism in the Far East*. (New York: Henry Holt, 1929.)

[278]   Paret, Peter. *Makers of Modern Strategy*. (Princeton, NJ: Princeton University Press, 1986.)

[279]   Parkinson, Roger. *Clausewitz*. (New York: Stein and Day, 1979.)

[280]   Parrish, Robert, and N. A. Andreacchio. *Schwarzkopf*. (New York: Bantam, 1991.)

[281]   Patai, Raphael. *The Arab Mind*. (New York: Scribner, 1983.)

[282]   Pauling, Linus, E. Laszlo, and J. Y. Yoo. *World Encyclopedia of Peace*. (Oxford, England: Pergamon Press, 1986.)

[283]   Payne, Keith B. *Missile Defense in the 21st Century*. (Boulder, CO: Westview Press, 1991.)

[284]   Payne, Samuel B., Jr. *The Conduct of War*. (New York: Basil Blackwell, 1989.)

[285]   Peeters, Peter. *Can We Avoid a Third World War Around 2010?* (London: The Macmillan Press, 1979.)

[286]   Pepper, David, and Alan Jenkins, eds. *The Geography of War*. (New York: Basil Blackwell, 1985.)

[287]   Perlmutter, Amos. *The Military and Politics in Modern Times*. (New Haven, CT: Yale University Press, 1977.)

[288]   Peters, Cynthia. *Collateral Damage*. (Boston: South End Press, 1992.)

[289]   Petre, F. Loraine. *Napoleon at War*. (New York: Hippocrene, 1984.)

[290]   Pierre, Andrew J., ed. *The Conventional Defense of Europe*. (New York: Council on Foreign Relations, 1986.)

[291]   Pipes, Daniel. *In the Path of God: Islam and Political Power*. (New York: Basic, 1983.)

[292]   Pisani, Edgard. *La Region . . . pour quoi faire?* (Paris: Calmann-Levy, 1969.)

[293]   Poggi, Gianfranco. *The Development of the Modern State*. (Stanford, CA: Stanford University Press, 1978.)

[294]   Polanyi, Karl. *The Great Transformation*. (Boston: Beacon Press, 1957.)

[295]   Polenberg, Richard. *War and Society: The United States, 1941–1945*. (New York: J. B. Lippincott, 1972.)

[296]   Polk, William R. *The Arab World Today*. (Cambridge, MA: Harvard University Press, 1991.)

[297]   Polmar, Norman, ed. *Soviet Naval Developments*. (Annapolis, MD: The Nautical and Aviation Publishing Company of America, 1979.)

[298]   Polmar, Norman, and Thomas B. Allen. *World War II: America at War, 1941–1945*. (New York: Random House, 1991.)

[299]   Price, Alfred. *Air Battle Central Europe*. (New York: The Free Press, 1987.)

[300]   Prigogine, Ilya. *Order Out of Chaos*. (New York: Bantam, 1984.)

[301]   Pujol-Davila, Jose. *Sistema y Poder Geopolitico.* (Buenos Aires, 1985.)

[302]   Quarrie, Bruce. *Special Forces.* (London: Apple Press, 1990.)

[303]   Read, James Morgan. *Atrocity Propaganda, 1914–1919.* (New Haven, CT: Yale University Press, 1941.)

[304]   Reese, Mary Ellen. *General Reinhard Gehlen: The C.I.A. Connection.* (Fairfax, VA: George Mason University Press, 1990.)

[305]   Renner, Michael. *Critical Juncture: The Future of Peacekeeping.* (Washington, D.C.: Worldwatch Paper, 1993.)

[306]   ———. *Swords Into Plowshares: Converting to a Peace Economy.* (Washington, D.C.: Worldwatch Paper, 1990.)

[307]   Renninger, John P. *The Future Role of the United Nations in an Interdependent World.* (Boston: Martinus Nijhoff, 1984.)

[308]   Rheingold, Howard. *Virtual Reality.* (New York: Summit, 1991.)

[309]   Rice, Edward E. *Wars of the Third Kind.* (Berkeley, CA: University of California Press, 1988.)

[310]   Richelson, Jeffrey T. *Foreign Intelligence Organizations.* (Cambridge, MA: Ballinger, 1988.)

[311]   ———. *The U.S. Intelligence Community.* (Cambridge, MA: Ballinger, 1985.)

[312]   Rinaldi, Angela, ed. *Witness to War: Images from the Persian Gulf War.* (Los Angeles: Los Angeles Times, 1991.)

[313]   Rivers, Gayle. *The Specialist.* (New York: Stein and Day, 1985.)

[314]   Robertson, Eric. *The Japanese File.* (Singapore: Heinemann Asia, 1979.)

[315]   Rogers, Barbara, and Zdenek Cervenka. *The Nuclear Axis.* (New York: Times Books, 1978.)

[316]   Romjue, John L. *From Active Defense to Airland Battle: The Deployment of Army Doctrine, 1973–1982.* (Fort Monroe, VA: Historical Office — U.S. Army Training and Doctrine Command, 1984.)

[317]   Rosecrance, Richard. *The Rise of the Trading State.* (New York: Basic, 1986.)

[318]   Rothschild, J. H. *Tomorrow's Weapons.* (New York: McGraw-Hill, 1964.)

[319]   Rustow, Alexander. *Freedom and Domination.* (Princeton, NJ: Princeton University Press, 1980.)

[320]   Safran, Nadav. *Israel: The Embattled Ally.* (Cambridge, MA: The Belknap Press, 1978.)

[321]   Sakaiya, Taichi. *The Knowledge-Value Revolution.* (New York: Kodansha International, 1991.)

[322]   Sallagar, Frederick M. *The Road to Total War.* (New York: Van Nostrand Rheinhold, 1969.)

[323]   Sampson, Anthony. *The Arms Bazaar.* (London: Coronet, 1983.)

[324]   Sanford, Barbara, ed. *Peacemaking.* (New York: Bantam, 1976.

[325]   Sardar, Zauddin, S. Z. Abedin, and M. A. Anees. *Christian-Muslim Relations: Yesterday, Today and Tomorrow.* (London: Grey Seal, 1991.)

[326]   Schevill, Ferdinand. *A History of Europe.* (New York: Harcourt, Brace, 1938.)

[327]   Schlosstein, Steven. *Asia's New Little Dragons.* (Chicago: Contemporary, 1991.)

[328]   Schoenbrun, David. *Soldiers of the Night.* (New York: E. P. Dutton, 1980.)

[329] Schreiber, Jan. *The Ultimate Weapon: Terrorists and World Order.* (New York: William Morrow, 1978.)

[330] Schwarzkopf, H. Norman, and Peter Petre. *It Doesn't Take a Hero.* (New York: Bantam, 1992.)

[331] Schweizer, Peter. *Friendly Spies.* (New York: Atlantic Monthly Press, 1993.)

[332] Scowcroft, Brent, ed. *Military Service in the United States.* (Englewood Cliffs, NJ: Prentice-Hall, 1982.)

[333] Seaton, Albert. *The German Army: 1933–1945.* (New York: Meridian, 1985.)

[334] Seaton, Albert, and Joan Seaton. *The Soviet Army: 1918 to the Present.* (New York: New American Library, 1987.)

[335] Seth, Ronald. *Secret Servants.* (New York: Farrar, Straus and Cudahy, 1957.)

[336] Seward, Desmond. *Metternich, The First European.* (New York: Viking, 1991.)

[337] ———. *Napoleon and Hitler.* (New York: Viking, 1989.)

[338] Shafer, Boyd C. *Faces of Nationalism.* (New York: Harvest, 1972.)

[339] Shaker, Steven M., and Alan R. Wise. *War Without Men: Vol. II, Future Warfare Series.* (Washington, D.C.: Pergamon-Brassey's, 1988.)

[340] Sharp, Gene. *Civilian-Based Defense.* (Princeton, NJ: Princeton University Press, 1990.)

[341] ———. *The Politics of Nonviolent Action: Part I-III.* (Boston: Porter Sargent, 1984–85.)

[342] Shaw, Martin. *Post-Military Society.* (Philadelphia: Temple University Press, 1991.)

[343] Shawcross, William. *The Quality of Mercy.* (New York: Simon and Schuster, 1984.)

[344] Sherwood, Robert M. *Intellectual Property and Economic Development.* (Boulder, CO: Westview Press, 1990.)

[345] Shultz, Richard H., and Roy Godson. *The Strategy of Soviet Information.* (New York: Berkley, 1986.)

[346] Simpkin, Richard. *Race to the Swift.* (New York: Brassey's Defence Publishers, 1985.)

[347] Singer, J. David. *Explaining War.* (Beverly Hills, CA: Sage Publications, 1979.)

[348] Singlaub, John, and Malcolm McConnell. *Hazardous Duty.* (New York: Summit, 1991.)

[349] Smith, Perry. *How CNN Fought the War.* (New York: Birch Lane Press, 1991.)

[350] Speiser, Stuart M. *How to End the Nuclear Nightmare.* (Croton-on-Hudson, NY: North River Press, 1984.)

[351] Stableford, Brian, and David Langford. *The Third Millenium: A History of the World AD 2000–3000.* (New York: Knopf, 1985.)

[352] Stanford, Barbara, ed. *Peacemaking.* (New York: Bantam, 1976.)

[353] Starr, Chester G. *The Influence of Sea Power on Ancient History.* (New York: Oxford University Press, 1989.)

[354] Stephens, Mitchell. *A History of News.* (New York: Viking, 1988.)

[355] Sterling, Claire. *The Terror Network.* (New York: Berkley, 1982.)

[356]   Stine, G. Harry. *Confrontation in Space.* (Englewood Cliffs, NJ: Prentice-Hall, 1981.)

[357]   Stoessinger, John G. *Why Nations Go to War.* (New York: St. Martin's, 1974.)

[358]   Strachey, John. *The End of Empire.* (New York: Random House, 1960.)

[359]   ———. *On the Prevention of War.* (New York: St. Martin's, 1963.)

[360]   Strassmann, Paul. *The Business Value of Computers.* (New Canaan, CT: The Information Economics Press, 1990.)

[361]   Strauss, Barry S., and Josiah Ober. *The Anatomy of Error.* (New York: St. Martin's, 1990.)

[362]   Strausz-Hupe, Robert. *The Balance of Tomorrow.* (New York: G. P. Putnam's Sons, 1945.)

[363]   Sulzberger, C. L. *World War II.* (New York: American Heritage Press, 1970.)

[364]   Summers, Col. Harry G., Jr. *On Strategy: A Critical Analysis of the Vietnam War.* (New York: Dell, 1982.)

[365]   Suter, Keith D. *A New International Order.* (Australia: World Association of World Federalists, 1981.)

[366]   ———. *Reshaping the Global Agenda: The U.N. at 40.* (Sydney: U.N. Association of Australia, 1986.)

[367]   Suvorov, Viktor. *Inside the Aquarium.* (New York: Berkley, 1986.)

[368]   ———. *Inside the Soviet Army.* (London: Hamish Hamilton Ltd., 1982.)

[369]   ———. *Inside Soviet Military Intelligence.* (New York: Berkley, 1984.)

[370]   Taber, Robert. *The War of the Flea: Guerilla Warfare Theory and Practice.* (London: Paladin, 1970.)

[371]   Taylor, Philip M. *Munitions of the Mind.* (Wellingborough, England: Patrick Stephens, 1990.)

[372]   ———. *War and the Media.* (Manchester: Manchester University Press, 1992.)

[373]   Taylor, William J., Jr., and Steven A. Maaranen, eds. *The Future of Conflict in the 1980's.* (Lexington, MA: Lexington, 1984.)

[374]   Tefft, Stanton K. *Secrecy.* (New York: Human Sciences Press, 1980.)

[375]   Thayer, George. *The War Business.* (New York: Discus, 1970.)

[376]   Thurow, Lester. *Head to Head.* (New York: William Morrow, 1992.)

[377]   Timasheff, Nicholas S. *War and Revolution.* (New York: Sheed and Ward, 1965.)

[378]   Toffler, Alvin, and Heidi Toffler. *Future Shock.* (New York: Bantam, 1970.)

[379]   ———. *Powershift.* (New York: Bantam, 1990.)

[380]   ———. *Previews & Premises.* (New York: William Morrow, 1983.)

[381]   ———. *The Third Wave.* (New York: Bantam, 1980.)

[382]   Trotter, W. *Instincts of the Herd in Peace and War.* (London: T. Fisher Unwin, 1917.)

[383]   Tuchman, Barbara W. *A Distant Mirror.* (New York: Knopf, 1978.)

[384]   Tuck, Jay. *High-Tech Espionage.* (London: Sidgwick and Jackson, 1986.)

[385]   Turner, Stansfield. *Secrecy and Democracy.* (Boston: Houghton Mifflin, 1985.)

[386]   ———. *Terrorism and Democracy.* (Boston: Houghton Mifflin, 1991.)

[387] Tzu, Sun (Griffith, Samuel B., trans.) *The Art of War.* (New York: Oxford University Press, 1963.)

[388] Ury, William L. *Beyond the Hotline.* (Boston: Houghton Mifflin, 1985.)

[389] Vagts, Alfred. *A History of Militarism: Civilian and Military.* (New York: Meridian, 1959.)

[390] Walzer, Michael. *Just and Unjust Wars.* (New York: Basic, 1992.)

[391] Warden, John A., III. *The Air Campaign: Planning for Combat.* (Washington, D.C.: Pergamon-Brassey's, 1989.)

[392] Watson, Peter. *War on the Mind.* (New York: Basic, 1978.)

[393] Weizsacker, Carl Friedrich von. *The Politics of Peril: Economics and the Prevention of War.* (New York: The Seabury Press, 1978.)

[394] Wells, H. G. *War and the Future.* (New York: Cassell, 1917.)

[395] Williams, Glyndwr. *The Expansion of Europe in the Eighteeth Century.* (New York: Walker, 1967.)

[396] Wilson, Andrew. *The Bomb and the Computer.* (New York: Delacorte Press, 1968.)

[397] Wittfogel, Karl A. *Oriental Despotism: A Comparative Study of Total Power.* (New Haven, CT: Yale University Press, 1964.)

[398] Woodruff, William. *The Struggle for World Power: 1500–1980.* (New York: St. Martin's, 1981.)

[399] Woodward, David. *Armies of the World: 1854–1914.* (New York: Putnam, 1978.)

[400] Worrall, R. L. *Footsteps of Warfare.* (London: Peter Davies, 1936.)

[401] Yarmolinsky, Adam. *The Military Establishment.* (New York: Harper & Row, 1971.)

[402] Yeselson, Abraham, and Anthony Gaglione. *A Dangerous Place: The United Nations as a Weapon in World Politics.* (New York: Viking, 1974.)

[403] Zhukov, Y. M. *The Rise and Fall of the Gumbatsu.* (Moscow: Progress Press, 1975.)

[404] *The Airland Battle and Corps 86: Tradoc Pamphlet 525-5.* (Fort Monroe, VA: U.S. Army Operational Concepts, March 21, 1981.)

[405] *The Annual Report for 1990.* (Austria: International Atomic Energy Agency, 1991.)

[406] *Common Security: A Blueprint for Survival.* (New York: Simon and Schuster, 1982.)

[407] *Conduct of the Persian Gulf War:* D.O.D. Final Report to Congress. (Washington, D.C.: U.S. Government Printing Office, 1992.)

[408] *Essays on Strategy* — 1984 Joint Chiefs of Staff Essay Competition Selections. (Washington, D.C.: National Defense University Press, 1985.)

[409] *From Semaphore to Satellite.* (Geneva: International Telecommunications Union, 1965.)

[410] *U.S. Army Field Manual (FM) 100–5, Operations, August 20, 1982.*

[411] *U.S. Army Field Manual (FM) 100–5, Operations, June 14, 1993.*

INDEX

Damascus, 49
DARPA. *See* Defense Advanced Research
   Project Agency
Data. *See* Knowledge
Data banks, 59, 139, 202, 249
Davie, Maurice R., 33
Davis, Edward E., 111
Davis, Richard Harding, 172
de Briganti, Giovanni, 260
de Czege, Huba Wass, 53
de Marenches, Count, 157
de Seversky, Alexander, 30
Death Pays a Dividend, 183
Decentralization, 10, 35, 77–78, 118, 156,
   161, 169, 207
Decision-making, 63, 160
De-coupling rich and poor, 22–23. *See also*
   Revolt of the rich
Deep battle, 51, 53–55, 67, 68; defined,
   51, 53
Defence Ministry (U.K.), 102
Defense Advanced Research Project Agency,
   115
Defense industry, 72, 108, 134, 142, 148, 179–
   189, 234. *See also* Weapons; *names of
   companies*
Defense Intelligence Agency, 155, 163
Defense News, 108, 115, 148, 165, 187
Defense Nuclear Agency, 53
Definitions: anti-war, 4; bisected global
   power, 21; civilization, 21, 256;
   civilizations, clash of, 18; First Wave
   civilization, 18–19; knowledge (*see*
   Knowledge); master conflict, 20; Second
   Wave civilization, 19; Third Wave
   civilization, 21–22; trisected global power,
   21, 219; war, 33–34; war-form, 81–82
Dehaven, John, 262
Delta Force, 91
De-massification, concept, 72; of destruction,
   67, 72–73; of family system, 22; of finance,
   90; of intelligence, 118, 156; of markets, 20,
   59–60, 89–90, 185, 246; of media, 22, 60,
   171; of military, 72; of production, 22, 51,
   59–60, 72, 83, 188; of threats; 90; of
   weapons, 72, 73, 192
Democracy, 11, 15, 19, 96, 133–134, 168, 172,
   189, 209
Denmark, 92, 250
Denver, 60, 150
DePuy, William E., 50–51, 53
Descartes, René, 38
Desert Shield, 79
Desert Storm, 54, 69–70, 74–75, 111
Destruction, mass. *See* Mass destruction
Deterrence, 4, 84, 116, 129, 198, 202. *See also*
   Mutually Assured Destruction
Development, uneven, 20

DEW Line (Distant Early Warning system),
   200
DIA. *See* Defense Intelligence Agency
Dictatorship, 44, 82, 155, 195
Digby, James F., 73
Diodorus Siculus, 30
Diplomacy, 15, 85, 127, 134, 155–156, 161,
   166, 170, 172, 191, 210, 228
Direct broadcast satellite. *See* Media;
   Satellites
Disarmament. *See* Arms control and
   disarmament
Disaster, 56, 91, 155, 205, 208, 228, 236, 251
Disease. *See* Health
Diseconomies of scale. *See* Scale and scale
   effects
Dissipative structures, 250
Division of labor, 40, 83
DMSO, 131
DNA, 56, 96, 122
Doctrine, military, 10–11, 29, 32, 39, 41–43,
   44–56, 66, 68–69, 73–74, 93, 127–128, 132,
   134–135, 139–152, 175; AirLand Battle,
   44–56, 66, 68–69, 74, 132, 134–135, 140,
   175; anti-war, 134; non-lethal, 127–128,
   132; Soviet, 52. *See also* Knowledge
   strategy
Douhet, Giulio, 68
Dr. Strangelove (film), 116
Draft. *See* Conscription
Dragon (anti-tank weapon), 65
Dresden, 41, 245
Drexler, K. Eric, 261
Dror, Yehezkel, 250
Drugs, *See* Narcotics
Dual-use technology, 112, 185, 204, 232. *See
   also* Civilianization
Dublin, 147
Duck Soup (film), 24
Dueling, 15, 224
Dukakis, Michael, 208
Duke, David, 236–237

East Bloc, 46
Eastern Europe. *See* Central Europe
Ecology. *See* Environment and
   environmentalism; War, ecological
Economic warfare, 16. *See also* Geo-
   economics, theory of
Economies of scale. *See* Scale and scale
   effects
Economist, 197
Economy, 51; brain-force (*see* Economy,
   Third Wave); brute-force (*see* Second Wave
   civilization); capital (*see* Capital); factors of
   production, 59; geo-economic theory (*see*
   Geo-economics); knowledge base of, 4–5,
   57–63; military parallels to, 10, 16–17,